MAKING MATHS COUNT

EXPLORING MATHS CONCEPTS IN REAL-WORLD CONTEXTS

PETER MAHER

OXFORD

UNIVERSITY PRESS

OXFORD
UNIVERSITY PRESS

Oxford University Press is a department of the University of Oxford.
It furthers the University's objective of excellence in research,
scholarship, and education by publishing worldwide. Oxford is a registered
trademark of Oxford University Press in the UK and in certain other countries.

Published in Australia by
Oxford University Press
Level 8, 737 Bourke Street, Docklands, Victoria 3008, Australia.

A catalogue record for this
book is available from the
National Library of Australia

ISBN 9780190338565

Reproduction and communication for educational purposes
The Australian *Copyright Act 1968* (the Act) allows educational institutions that
are covered by remuneration arrangements with Copyright Agency to reproduce
and communicate certain material for educational purposes. For more information,
see copyright.com.au.

Edited by Tom Smallman
Produced by Newgen KnowledgeWorks Pvt. Ltd., Chennai, India
Printed in China by Sheck Wah Tong Printing Press Ltd.

Disclaimer
Aboriginal and Torres Strait Islander peoples are advised that this publication may
include images or names of people now deceased.

*Links to third party websites are provided by Oxford in good faith and for information only.
Oxford disclaims any responsibility for the materials contained in any third party website
referenced in this work.*

CONTENTS

OXFORD UNIVERSITY PRESS

INTRODUCTION

The teaching of mathematics has undergone fundamental change in recent times. In fact, today's effective mathematics classrooms and teaching programs bear little resemblance to those of a generation ago. At the school where I currently teach, early in the school year we invite parents and carers to attend lessons at which these fundamental changes are demonstrated. Many participants are astounded at how students learn numeracy in school today.

The research of the Swiss psychologist Jean Piaget has shown us that the best way to cater for the learning needs of students is to offer activities that match their intellectual stage of development. Most primary students best abstract new mathematical concepts when they are presented to them in a manner that allows them to manipulate materials or see pictures, drawings or icons. To paraphrase Confucius, the best teachers of mathematics adhere to the adage: 'I hear and I know, I see and I learn, I do and I understand.'

The fact that we are all a part of a society that is changing at a breakneck pace has critical implications for educational institutions. One of the fundamental roles of schools is to enable its students to gain intellectual independence and thus be able to cope with life both inside and outside the classroom, as well as the demands of the workplace when their formal education is complete. No one can predict the jobs that will be available to our five-year-olds when they seek employment in the future. So, as teachers, the best thing we can do is to help develop the students under our care into flexible, creative thinkers. Teaching them strategies to problem solve – and regularly presenting tailor-made problems to solve – is essential to their development as successful, independent mathematicians.

British mathematician and psychologist Richard R. Skemp demonstrated that a student's understanding takes two basic forms – superficial, rote understanding and deep, applicable understanding. The best teachers of mathematics are not fully satisfied when their students respond with a correct answer; they are satisfied when their students respond with a correct answer and can explain why their answers are correct. A deep understanding of mathematical concepts leads to successful application and problem solving, the ability to find patterns – the fundamental building block of the subject – and long-term retention and numerical fluency.

ABOUT *MAKING MATHS COUNT*

The 17 units of work in *Making Maths Count* are based on encouraging deep understanding. They focus on two other keystones of effective mathematics education – relevance and differentiation.

Relevance

Mathematics is ubiquitous. If we truly believe that the role of schools is to prepare students for life beyond the classroom, the mathematics we teach must be relevant to students' everyday lives. Topics must be taught in a 'real life' context and should offer students a reason for learning. Engagement and motivation will almost certainly follow, leading to further learning. Each topic in *Making Maths Count* starts with a brainstorming session during which the student is asked to consider where the topic being studied has affected their lives or can be seen in the broader community. This session will give the teacher an indication of the depth of knowledge the students already have, as well as a sense of the varying ranges of conceptualisation within the class. Questions like 'Have you heard of the term decimal?', 'What does a decimal look like?', 'Where might you see a decimal being used?' and so on offer excellent learning opportunities for teachers and students alike. This session will invariably lead to different shared responses and an accompanying growth in mathematical vocabulary. The work that follows then shows the relevance of the topic to the 'real world'.

Differentiation

Each unit of work in *Making Maths Count* adopts the 'low threshold, high ceiling' approach. Gone are the days when the simple philosophy of 'I teach a class' was acceptable. We know that a class is made up of individuals with a wide range of radically different capabilities and personalities. It is not unusual within any given primary class to encounter students with vastly different mathematical capabilities. The units in *Making Maths Count* are structured in such a manner that every individual in the class will be able to find a way to engage with and answer questions at their own level. Each unit starts with a basic activity, often drawing upon a students' own experience, and then moves into more-demanding aspects increasing in difficulty. By the end of each unit, the student is challenged with problem-solving activities and applications that will enrich and extend their learning.

Some students in your class will only be able to complete the first topic, or the first few topics, in each unit, while others will be able to attempt and/or complete them all. The book is structured so that both the less able and the very able students in your class will be challenged and engaged. The tasks generally become more difficult as the student progresses through each unit. Peer coaching works wonderfully well in most classrooms – if possible, encourage your more able students to work with those who are less capable. Let these students be the teacher from time to time. The less able will enjoy a fresh approach from a peer, and the teaching will help crystallise the concept under review in the minds of the more capable.

Making Maths Count focuses on practical, engaging and differentiated tasks that are designed to show all primary students the joy and inherent beauty that can be found in the pursuit of mathematical inquiry.

HOW TO USE THIS BOOK

Making Maths Count was written in the period of remote learning in Melbourne during the COVID-19 pandemic in 2020, and the tasks were designed to be completed independently by students at home. It has been adapted here to offer teachers a curated collection of activities to support mathematical learning both in the classroom, or at home, and with individual students, or groups of students. The tasks can be used to support whole-classroom learning and whole-school maths programs, to encourage engagement in struggling students or extend more able students, as homework activities, or as whole-class/group/independent activities. All units are aligned with the Australian Curriculum: Mathematics.

ACTIVITIES

Activities within the unit are graded in difficulty from Task 1 to Task 10, so that students of all abilities are catered for, and each unit is based on a mathematical concept. Tasks are not prescribed by year level.

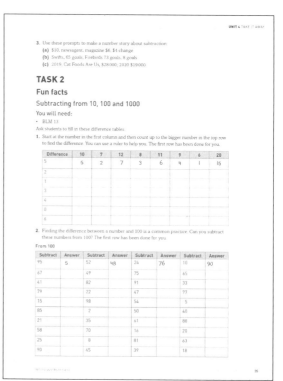

MATERIALS

Many tasks include a list of materials to allow teachers to prepare in advance. It is assumed that everyday classroom materials – such as paper, pencils, counters, Unifix or Multilink cubes and Base 10 materials – are readily available. However, many tasks also refer to materials commonly found in the home – such as food packaging or playing cards – because the more students can relate maths activities to their everyday lives, the more likely that mathematical concepts and understanding will resonate with them. Materials can be adapted to suit the school or home environment, depending on where the students are carrying out the tasks.

OXFORD UNIVERSITY PRESS

INSTRUCTIONS AND ANSWERS

Each unit includes a short introduction that gives ideas for brainstorming sessions or points of interest for discussion, relating the topic back to everyday experiences and general knowledge. Each task has clear, simple instructions and answers are provided at the end of each unit.

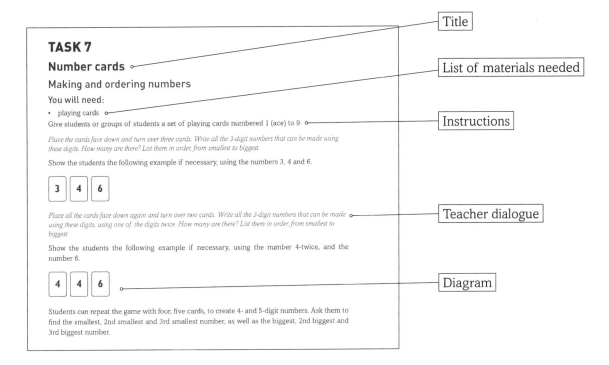

BLMs

All BLMs are available online at oxfordowl.com.au.

They include number cards, grids and templates, and the recording tables used in many of the tasks throughout the book.

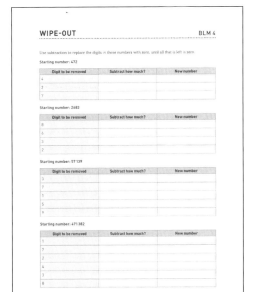

UNIT 1 – KNOW YOUR PLACE
Number and place value

Our system of counting and numeration is based on 10 different digits, from 0 to 9. This is called a Base 10 system. In the Base 10 system, the biggest digit possible in any place or column (if you are using a place value chart) is 9. The concept of basing numeration on 10 different digits, one for each finger, was developed in northern India in the 6th or 7th century and spread around the world by Arab traders on the Silk Road – hence the name 'Hindu-Arabic'. The system entered Europe in the early 13th century, thanks to Leonardo Fibonacci.

It can be strongly argued that place value and numeration are the cornerstones of mathematics. Without a good basic knowledge of the way that our Hindu-Arabic Base 10 numeration system works, students will be building future conceptualisation on an unstable, constantly shifting foundation of sand.

Place value should be one of the first topics taught each year, with a heavy reliance on the place value chart and as many different forms of concrete materials as possible:

Millions	Hundreds of thousands	Thousands	Hundreds	Tens	Ones
1 000 000s	100 000s	1000s	100s	10s	1s

Often, knowing the exact size of a number is not necessary. This is especially the case when we encounter large numbers. Numbers can get rounded up or down to make them easier to deal with, for convenience and accessibility. For example, if there are 523 students in a whole-school assembly, we might say there are 'about 520' or even 'about 500' students in the assembly. Brainstorm other situations where we encounter large numbers, and round them up or down with the students. For example, populations of cities, states and countries; crowd sizes at sports events; the number of students in a school or year level; and the costs of large or expensive purchases (e.g. houses, cars).

Make sure that students understand that numbers are usually rounded up if they are more than halfway to the next 10, 100 or 1000. For example, 75 gets rounded up to 80 because 75 is the 6th number in the 70s (70, 71, 72, 73, 74, 75, 76, 77, 78, 79). Any number from 71 to 74 gets rounded down to 70. When you round a number to the nearest 100, 50 becomes the key. For example, 450 gets rounded up to 500. There are 50 numbers from 400 to 449 and 50 numbers from 450 to 499, so 450 is the start of the second half of numbers in the 400s and is rounded to 500. Any number below 450 is rounded down to 400.

TASK 1

Turn card challenge

Ranking numbers to 100

You will need:

- playing cards
- BLM 1

Give pairs or small groups of students a deck of playing cards and take out the jacks, queens, kings and jokers. Tell the students that in this game, the aces will stand for the digit 1 and the 10s cards will stand for the digit 0. Ask the students to shuffle the cards and put them face down in a pile.

Turn two cards over to create a number. The first card you turn over will be the digit in the tens place, and the second card you turn over will be the digit in the ones place. So, if you turn over a 6 then a 10, it will make 60. If you turn over a 10 then a 4, it will make 4, because a 10 card is a zero.

Ask the students to record the numbers they make as they go. Make sure they mix up the cards every time they have a go. Some examples are shown here.

Two-digit numbers										
90 to 99	92									
80 to 89										
70 to 79		74								
60 to 69										
50 to 59										
40 to 49			43							
30 to 39										
20 to 29										
10 to 19										
0 to 9										

Now write down the numbers you have made in order from smallest to biggest. What was the closest number to 100 that you made? How far away from 100 was this number?

TASK 2

The biggest number wins

Comparing 3-digit numbers

You will need:

- playing cards
- BLM 2

Give each pair of students a set of cards from ace to 10. Tell the students that in this game, the ace will stand for the digit '1' and the 10 will stand for the digit '0'. Each player starts with 50 points and the students take turns to go first. In each game, after the first player has made a number, the second player can risk up to 10 points of their total 50 points.

Shuffle the cards and choose three random cards. The first card you pick will be the digit in the hundreds place, the second card will be the digit in the tens place and the third card will be in the ones place. So, if you pick a 2, then a 6, then a 10, your number will be 260.

After the first player has made a number, the second player chooses cards to make a number. The player with the higher number wins the points risked, and the winner is the player with the greater number of points after 20 rounds. Some examples are shown here:

Game	Player 1 number	Player 2 risk	Player 2 number	Player 1 points	Player 2 points
1	260	10	139	60	40
2	145	20	398	40	60
3					
4					
5					
6					
7					
8					
9					
10					
11					
12					
13					
14					
15					
16					
17					
18					
19					
20					

Do you think it is better to go first or second in this game? Why?

TASK 3

Round it up, round it down

Rounding numbers to the nearest 10 or 100

You will need:

* playing cards
* BLM 3

Give pairs or groups of students a deck of cards, but first remove the picture cards (jacks, queens, kings and jokers). Tell the students that in this game, the ace will stand for the digit '1' and the 10 will stand for the digit '0'.

Put the cards face down on a table and choose two. The first card you pick will be the digit in the tens place and the second card you pick will be the digit in the ones place. So, if you pick a 4, then a 1, your number will be 41. Now round this number to the nearest 10. If you pick a 10, then a 7, your number will be 7. Now round this number to the nearest 10.

Number picked	Rounded to the nearest 10
41	40
7	10

Now choose three cards. The first card you pick will be the digit in the hundreds place, the second card will be the digit in the tens place and the third card will be the digit in the ones place. So, if you pick a 2, then a 4, then a 1, your number will be 241. Now round this number to the nearest 100. If you pick a 10, then a 7, then a 9, your number will be 79. Now round this number to the nearest 100. If you pick a 5, then a 10, then a 3, your number will be 503. Now round this number to the nearest 100.

Number picked	Rounded to the nearest 100
241	200
79	100
503	500

Number picked	Rounded to the nearest 100

TASK 4

Wipe-out

Recognising the place of digits in a number

You will need:

- BLM 4

In this game, students replace digits in a number with zeros, one at a time, until all that is left is zero. However, this can only be done by using subtraction. For example, take the number 358: subtract 50 to get 308, then subtract 8, leaving 300. Finally, subtract 300, to get 0. Alternatively, subtract 300 to get 58, then subtract 8 to get 50, and then subtract 50 to get 0.

Starting number: 358

Digit to be removed	Subtract how much?	Number left
5	50	308
8	8	300
3	300	0

Use subtraction to replace the digits in these numbers with zero, until all that is left is zero.

Starting number: 472

Digit to be removed	Subtract how much?	New number
4		
2		
7		

Starting number: 2683

Digit to be removed	Subtract how much?	New number
8		
6		
3		
2		

Starting number: 57 139

Digit to be removed	Subtract how much?	New number
3		
7		
1		
5		
9		

Starting number: 471 382

Digit to be removed	Subtract how much?	New number
1		
7		
2		
4		
3		
8		

TASK 5

Think big

Creating large numbers

You will need:

- playing cards
- BLM 5

Give each student or groups of students a deck of cards, but first remove the picture cards (jacks, queens, kings and jokers). Tell the students that in this game, the ace will stand for the digit '1' and the 10 will stand for the digit '0', and that the aim is to make the biggest possible number.

First make a 3-digit number. Shuffle the cards and choose one. Decide where to place this digit so that you will end up with the biggest possible number. Return the card to the deck, shuffle the cards and repeat this until your number is complete. Did you end up with the biggest possible number? Can you name this number?

Ask students to repeat the game for 4-, 5-, 6-, 7-, 8-, 9- and 10-digit numbers.

Write your answer and the biggest possible answer for the following:

OXFORD UNIVERSITY PRESS

- *Three digits – Hundreds*
- *Four digits – Thousands*
- *Five digits – Tens of thousands*
- *Six digits – Hundreds of thousands*
- *Seven digits – Millions*
- *Eight digits – Tens of millions*
- *Nine digits – Hundreds of millions*
- *Ten digits – Billions*

TASK 6

Population count

Ordering and rounding large numbers

You will need:

- BLM 6

Tell students that these tables show the populations of cities and states around Australia in 2020. For each population, pronounce it using the place value chart: *units, tens, hundreds of ones, units, tens, hundreds of thousands, units, tens, hundreds of millions.* Ask students to find the value of a given digit and then round the number to the nearest 1000.

Place	Population	Value of the 4	Population rounded to nearest 1000
Broome	14 503		
Whyalla	21 478		
Goulburn	24 382		
Tamworth	43 251		
Mildura	52 314		
Shepparton	52 549		
Bundaberg	71 554		
Cairns	155 340		
Geelong	282 412		
Melbourne	4 969 305		

Which two cities have the closest populations?

Place	Population	Value of the 2	Population rounded to nearest 100 000
Northern Territory	246 143		
Adelaide	1 352 504		

Place	Population	Value of the 2	Population rounded to nearest 100 000
Brisbane	2 475 680		
Western Australia	2 663 561		
Sydney	4 966 826		
Queensland	5 255 035		
Victoria	6 666 862		
New South Wales	8 082 508		
Australia	25 704 340		

Which two states in this table are closest in population?

TASK 7

Number cards

Making and ordering numbers

You will need:

- playing cards

Give students or groups of students a set of playing cards numbered 1 (ace) to 9.

Place the cards face down and turn over three cards. Write all the 3-digit numbers that can be made using these digits. How many are there? List them in order, from smallest to biggest.

Show the students the following example if necessary, using the numbers 3, 4 and 6.

Place all the cards face down again and turn over two cards. Write all the 3-digit numbers that can be made using these digits, using one of the digits twice. How many are there? List them in order, from smallest to biggest.

Show the students the following example if necessary, using the number 4-twice, and the number 6.

Students can repeat the game with four, five cards, to create 4- and 5-digit numbers. Ask them to find the smallest, 2nd smallest and 3rd smallest number, as well as the biggest, 2nd biggest and 3rd biggest number.

OXFORD UNIVERSITY PRESS

Place the cards face down and turn over six cards. How many 6-digit numbers do you think can be made using these 6 different digits?

Place the cards face down and turn over nine cards. How many 9-digit numbers do you think can be made using these 9 different digits? What is the biggest possible number you could make? And the smallest?

TASK 8

Binary numbers

Using place value with Base 2

You will need:

- BLM 7

Explain to students that while we normally use a Base 10 number system, computer programmers use a Base 2, or binary, system. In a Base 2 system, the only digits that you can ever have in a place or column (if you are using a place value chart) are 0 and 1.

A binary place value chart looks like this:

2048	1024	512	256	128	64	32	16	8	4	2	1
2 × 2 × 2 × 2 × 2 × 2 × 2 × 2 × 2 × 2 × 2	2 × 2 × 2 × 2 × 2 × 2 × 2 × 2 × 2 × 2	2 × 2 × 2 × 2 × 2 × 2 × 2 × 2 × 2	2 × 2 × 2 × 2 × 2 × 2 × 2 × 2	2 × 2 × 2 × 2 × 2 × 2 × 2	2 × 2 × 2 × 2 × 2 × 2	2 × 2 × 2 × 2 × 2	2 × 2 × 2 × 2	2 × 2 × 2	2 × 2	2	1

So, if we want to make the number 22, we need 16 + 4 + 2 shown as:

2048	1024	512	256	128	64	32	16	8	4	2	1
							1	0	1	1	0

The binary number for 22 is 10110 (one, zero, one, one, zero).

Change these Base 10 numbers into binary:

	2048	1024	512	256	128	64	32	16	8	4	2	1
9												
37												
131												
305												
1100												
2500												

Change these binary numbers back into Base 10:

Number	2048	1024	512	256	128	64	32	16	8	4	2	1
									1	1	1	1
								1	0	0	1	1
							1	1	1	1	1	1
					1	1	0	0	0	0	0	1
		1	0	1	0	1	0	1	0	1	0	1
		1	1	1	1	1	1	0	0	1	0	0

TASK 9

Problems to solve

Applying place value to unfamiliar situations

You will need:

- BLM 7

Encourage students to try these challenging place value problems:

1. What is the biggest 5-digit number that, when rounded to the nearest 10, equals 34 670?

2. You have a million dollars. You buy the following items:

Magazine $10.10
Racing bike $1000
Sports car $100 000
How much money would you have left?

3. Can you find the mystery number?

I am a 6-digit number. When rounded to the nearest 100 000, I equal 800 000. My hundreds of thousands place is a prime number. My hundreds place is exactly 1000 times smaller than my hundreds of thousands place. Three of my digits are odd and three of my digits are even. My ones place is the second perfect square. My tens place is the second prime number. My ten thousands and thousands digits sum to 10. Only two of my digits are repeated.
What number am I?

4. Here is a Base 3 place value chart. In a Base 3 system, you can only use the digits 0, 1 and 2 when creating numbers.

In our Base 10 number system, what would the Base 3 number of 210 220 equal?

Write the year 2020 in Base 3.

2187	729	243	81	27	9	3	1
3 × 3 × 3 x 3 × 3 × 3 x 3	3 × 3 × 3 x 3 × 3 × 3	3 × 3 × 3 x 3	3 × 3 × 3 x 3	3 × 3 × 3	3 × 3	3	1

ANSWERS

Task 1 Turn card challenge

Answers will vary.

Task 2 The biggest number wins

Answers will vary.

It is better to go second in this game because you can use chance to help work out the size of the risk and your chance of winning.

Task 3 Round it up, round it down

Answers will vary.

Two-digit numbers ending in 1, 2, 3 or 4 should be rounded down to a tens place with a zero on the end. Two-digit numbers ending in 5, 6, 7, 8 or 9 should be rounded up to a tens place with a zero on the end.

Three-digit numbers ending in 01 to 49 should be rounded down to a hundreds place with two zeros on the end. Three-digit numbers ending in 50 to 99 should be rounded up to the next hundreds place with two zeros on the end.

Task 4 Wipe-out

Starting number: 472

Digit to be removed	Subtract how much?	New number
4	400	72
2	2	70
7	70	0

Starting number: 2683

Digit to be removed	Subtract how much?	New number
8	80	2603
6	600	2003
3	3	2000
2	2000	0

Starting number: 57139

Digit to be removed	Subtract how much?	New number
3	30	57109
7	7000	50109
1	100	50009
5	50000	9
9	9	0

Starting number: 471382

Digit to be removed	Subtract how much?	New number
1	1000	470382
7	70000	400382
2	2	400380
4	400000	380
3	300	80
8	80	0

Task 5 Think big

Answers will vary.

Encourage students to correctly pronounce each answer using the key of 'units, tens, hundreds of ones, units, tens, hundreds of thousands, units, tens, hundreds of millions, units, tens, hundreds of billions'.

Task 6 Population count

Place	Population	Value of the 4	Population rounded to nearest 1000
Broome	14503	4000	15000
Whyalla	21478	400	21000
Goulburn	24382	4000	24000
Tamworth	43251	40000	43000
Mildura	52314	4	52000
Shepparton	52549	40	53000
Bundaberg	71554	4	72000
Cairns	155340	40	155000
Geelong	282412	400	282000
Melbourne	4969305	4000000	4969000

Shepparton and Mildura have the closest populations.

Place	Population	Value of the 2	Population rounded to nearest 100 000
Northern Territory	246 143	200 000	200 000
Adelaide	1 352 504	2000	1 400 000
Brisbane	2 475 680	2 000 000	2 500 000
Western Australia	2 663 561	2 000 000	2 700 000
Sydney	4 966 826	20	5 000 000
Queensland	5 255 035	200 000	5 300 000
Victoria	6 666 862	2	6 700 000
New South Wales	8 082 508	2000	8 100 000
Australia	25 704 340	20 000 000	25 700 000

Queensland and Victoria are the states closest in population. Note that there are both states and cities in this table.

Task 7 Number cards

Answers will vary.

The number of possible answers follows factorials. This means the answer is calculated by multiplying together the number of digits by every number less than itself.

2 different digits: $2 \times 1 = 2$; 3 different digits: $3 \times 2 \times 1 = 6$; 4 different digits: $4 \times 3 \times 2 \times 1 = 24$;

5 different digits: $5 \times 4 \times 3 \times 2 \times 1 = 120$; 6 different digits: $6 \times 5 \times 4 \times 3 \times 2 \times 1 = 720$;

9 different digits: $9 \times 8 \times 7 \times 6 \times 5 \times 4 \times 3 \times 2 \times 1 = 362 880$.

The biggest possible number is 987 654 321 and the smallest possible is 123 456 789.

Task 8 Binary numbers

Base 10 to binary:

9 = 1001, 37 = 100101, 131 = 10000011, 305 = 100110001, 1100 = 10001001100, 2500 = 100111000100.

Binary to Base 10:

1111 = 15, 10011= 19, 111111 = 63, 11000001 = 193,

10101010101 = 1365, 1111100100 = 2020.

Task 9 Problems to solve

1. 34 674
2. $898 989.90
3. 782 734
4. 210 220 would equal $243 + 243 + 81 + 9 + 9 + 3 + 3 = 591$; and 2020 in Base 3 would equal 2 202 211.

UNIT 2 – CODE CRACKER
Sequences and series

Mathematics is based on patterns and connections. In fact, the brightest mathematicians in the class can best be identified by their ability to see patterns and to apply these patterns to different situations. Patterns or sequences can contain numbers, letters or symbols. They are defined by a rule (what is happening in each pattern) and by each term (each element of the pattern). Thus, the rule of the sequence 2, 4, 6, 8, ... is +2 or even numbers, and its 6th term would be 12. Knowing the rule and terms of a pattern or sequence will enable students to find missing terms. Understanding that even an equation is a type of pattern (with a mathematical balance that what is to the left of the equal sign is the same as what is to the right of the equal sign) demystifies the concept very well and can lead on to some quite sophisticated mathematics.

During the brainstorming session, think about patterns used to create or crack secret codes and encrypted computer programs. Secret codes have been used for centuries to disguise important messages or information. They are basically mathematical sequences, or variations on sequences.

Some of the most challenging codes are those where letters of the alphabet are shifted to disguise a message. Can you quickly decode the word IFMMP? It looks like nonsense until you move the letters back one spot in the alphabet and suddenly HELLO is revealed. One of the starting points to becoming a good code cracker and creator is to understand how often we use the letters in the English language.

As well as using numbers and letters, codes can be created by using symbols. The famous fictional detective Sherlock Holmes broke a symbol code in the story *The Adventure of the Dancing Men*. In this tale, Holmes used his knowledge of alphabet frequency to decipher a letter made up of stick figures. The most commonly used stick figure ended up standing for the letter E, the next most commonly used stick figure stood for the letter T and so on.

TASK 1

What comes next?

Finding missing terms in sequences

What number comes next in each of these sequences? Is there a pattern?

1, 2, 3, 4, 5 ...

2, 4, 6, 8, 10 ...

1, 4, 7, 10, 13 ...

5, 10, 15, 20, 25 ...

6, 16, 26, 36, 46 ...

11, 10, 9, 8, 7 ...

OXFORD UNIVERSITY PRESS

22, 20, 18, 16, 14 …
15, 12, 9, 6, 3 …
31, 27, 23, 19, 15 …
57, 47, 37, 27, 17 …
1, 2, 4, 8, 16 …
2, 20, 200 …
1, 3, 9 …
2, 10, 50 …
1, 4, 16 …
40, 20, 10 …
16, 8, 4 …
1000, 100, 10 …
48, 24, 12 …
125, 25, 5 …

TASK 2

Terms and rules

Recognising the structure of a sequence

Explain to the students that each number in a sequence is called a 'term'. Every sequence also has a 'rule'. For example, the 3rd term in the sequence 4, 8, 12, 16… is 12 and the rule is +4. Some sequences have a rule that can be described in more than one way. For example, in the sequence 2, 4, 6, 8… the rule could be either '+ 2' or 'the 2 × table' or 'even numbers'.

Find the rule and the 6th term for each of these sequences:

2, 5, 8, 11 …
1, 2, 4, 8 …
1, 2, 4, 7 …
32, 16, 8 …
2, 6, 18 …
100 000, 10 000, 1000 …
65, 59, 53, 47 …
60, 51, 42 …
25, 50, 75 …
100, 90, 81, 73 …

TASK 3

Letter patterns

Using sequences with letters

You will need:

• copy of the alphabet (BLM 73)

Explain to students that sequences do not only use numbers. Many sequences have letters of the alphabet as terms in their patterns.

Find the rule and the 8th term in each of these letter patterns:

A, B, C …
X, W, V …
A, C, E …
P, N, L …
C, F, I …
E, J, O …
Z, U, P …
A, C, F, J …
A, B, D, H …
G, V, K …

TASK 4

Double trouble

Using two-step sequences

Explain to students that some sequences contain a 'double pattern'. This is when two operations are performed in the rule: for example, adding 2 and then doubling, or dividing by 2 and then taking away 6.

Find the rule and the next term for each of these sequences with a double pattern:

2, 6, 14 …
1, 3, 7, 15 …
2, 5, 14 …
1, 7, 67 …
100, 48, 22 …
10, 6, 4 …
1, 9, 33 …
4, 6, 12, 30 …
10, 6, 4, 3 …
102, 50, 24 …

TASK 5

Abbreviation patterns

Using general knowledge to find missing terms

Explain to students that some sequences or patterns are created by using the first letter of a word. They can deal with cycles, like days of the week or months of the year, or collections like planets or chemical elements.

OXFORD UNIVERSITY PRESS

Use your general knowledge and logic to find the rule and the next term in each of these patterns:

M, T, W, T …

J, A, S, O …

O, T, T, F …

T, F, S, E …

T, T, T, F …

S, A, W, S …

T, T, F, S, E …

M, V, E, M …

E, S, W, N …

A, A, A, A, NA, SA …

TASK 6

Letter frequency

Using common letters

You will need:

- BLM 8

Explain to students that in the English language, some letters are used often and some hardly at all. The letter frequency in English is:

E T A O N I S R L H D C U F P M W Y B G V K Q X J Z

The first 11 letters are the ones we use very often: E T A O N I S R L H D. But we rarely use K Q X J Z.

Turn to any page of a book at random and record the letters you find. Soon you will see a pattern starting to form. Use BLM 8 to record your results until at least three of the columns are complete.

A	B	C	D	E	F	G	H	I	J	K	L	M	N	O	P	Q	R	S	T	U	V	W	X	Y	Z

Now list your letter count in order from the most found letter to the least found letter.

How closely did your letter count match the English letter frequency shown above?

TASK 7

Letter/Number codes

Using numbers that stand for letters

You will need:

• copy of the alphabet (BLM 73)

Remind students that in many codes and patterns, numbers stand for letters. For example, if a number appears often in a code, it might stand for a commonly used letter such as E or T or A – the three most common letters used in the English language. Sometimes the alphabet might be used backwards as well!

Think logically and consider alphabet order and frequency to crack these codes and find the secret word.

6 9 19 8

5 12 5 16 8 1 14 20

19 3 8 15 15 12

1 21 19 20 18 1 12 9 1

12 25 14 24

1 22 25 9 26

24 9 18 24 16 22 7

16 26 13 20 26 9 12 12

16 22 7 7 15 22

14 26 7 19 8

Can you crack the code to reveal the secret message?

25 15 21 – 1 18 5 – 22 5 18 25 – 7 15 15 4 – 1 20 – 3 18 1 3 11 9 14 7 – 3 15 4 5 19

TASK 8

Letter shift codes

Using letters that move

You will need:

• copy of the alphabet (BLM 73)

Remind students that in some codes and patterns, letters can be shifted forwards and backwards to make the pattern even harder to crack.

Use your knowledge of alphabet frequency to help crack these codes. The clues will help you.

Code: CBMMPPO (Clue: Toy.)

Code: CJSUIEBZ (Clue: Once a year.)

Code: BNLOTSDQ (Clue: IT.)

Code: QVQQZ (Clue: Pet.)

Code: JGNR! (Clue: Assistance.)

Code: YNNJC (Clue: Fruit.)

Code: DLOFIIX (Clue: Animal.)

OXFORD UNIVERSITY PRESS

Code: COXKZB (Clue: Country.)

Code: DVCCZ – JQWUG (Play space.)

Code: BMCFSU – FJOTUFJO – XBT – HSFBU – BU – DSBDLJOH – DPEFT

TASK 9

Symbol codes

Using symbols

You will need:

* BLM 9

In this symbol code, A = *, E = @, I = !, O = ^ and U = #. Can you figure out these words?

@ @ $ _____ (Clue: A slippery ray-finned fish.)

* $ @ _____ (Clue: A type of beer.)

$ * ` < _____ (Clue: Opposite of first.)

~ # ` < _____ (Clue: It often coats old pieces of iron.)

% * > & _____ (Clue: Your fingers are attached to it.)

This is your chance to be a detective like Sherlock Holmes. Decode these lines from a story about a sea captain who comes ashore at a strange town and is helped by a young boy.

"% @ $ $ ^"   ` \* ! &     < % @     ^ $ &     ` * ! $ ^ ~.   "+ * >   !
` < ~ ^ $ $     \* ~ ^ # > &     < % @     ` < ~ @ @ < `   / ! < %   ] ^ #?"

" ` # ~ @"     !     ` * ! &.   "! $! { @   ! >   < % ! `   < ^ / >."

"/ % * <   ! `     < % ! `   < ^ / >   + * $ $ @ &?"   * ` = @ &
< % @   ^ $ &   ` * ! $ ^ ~.

"! <   ! `     + \* $ $ @ &     / @ ` <   % * ~ | ^ # ~"   !   ` * ! &.

"* %,   < % @ ~ @' `     | # ~ ! @ &     < ~ @ \* ` # ~ @   % @ ~ @"
% @   ` * ! &.

TASK 10

Create a code

Making your own codes

By now, you are a code-cracking expert. You have solved sequences involving numbers, letters, cycles and symbols. Now use your skills and knowledge to become a code creator. Your task is to disguise a sentence, message or story using a sequence or pattern to make a code. Then pass on your word or message to a friend or an adult to try and decode. Kcul doog!

ANSWERS

Task 1 What comes next?

6, 12, 16, 30, 56, 6, 12, 0, 11, 7, 32, 2000, 27, 250, 64, 5, 2, 1, 6, 1.

Task 2 Terms and rules

Rule: + 3. 6th term: 17.

Rule: × 2. 6th term: 32.

Rule: + 1 more number each time. 6th term: 16.

Rule: Halving or dividing by 2. 6th term: 1.

Rule: × 3. 6th term: 486.

Rule: Dividing by 10. 6th term: 1.

Rule: − 6. 6th term: 35.

Rule: − 9. 6th term: 15.

Rule: + 25. 6th term: 150.

Rule: − 1 less number each time. 6th term: 60.

Task 3 Letter patterns

Rule: The alphabet. 8th term: H.

Rule: The alphabet backwards. 8th term: Q.

Rule: The alphabet starting with A and missing 1 letter. 8th term: O.

Rule: The alphabet backwards starting with P and missing 1 letter. 8th term: B.

Rule: Every 3rd letter of the alphabet. 8th term: X.

Rule: Every 5th letter of the alphabet, returning to the start of the alphabet. 8th term: N.

Rule: The alphabet backwards missing 4 letters each time. 8th term: Q.

Rule: The alphabet starting with A and missing 1 more letter each time. 8th term: I.

Rule: Doubling the place of the letters, 1st then 2nd then 4th then 8th. 8th term: This will be the '128th' letter of the alphabet, X.

Rule: − 10 letters starting at G. 8th term: H.

Task 4 Double trouble

Rule: + 1 then × 2. Next term: 30.

Rule: × 2 then + 1. Next term: 31.

Rule: × 3 then − 1. Next term: 41.

Rule: × 10 then − 3. Next term: 667.

Rule: ÷ 2 then − 2. Next term: 9.

Rule: ÷ 2 then + 1. Next term: 3.

Rule: + 2 then × 3. Next term: 105.

Rule: − 2 then × 3. Next term: 84.

Rule: + 2 then ÷ 2. Next term: 2.5.

Rule: − 2 then ÷ 2. Next term: 11.

Task 5 Abbreviation patterns

Rule: Days of the week. Next term: F.

Rule: Months of the year starting with July. Next term: N.

Rule: Counting numbers. Next term: F.

Rule: Even numbers. Next term: T.

Rule: Counting in tens. Next term: F.

Rule: The seasons. Next term: S.

Rule: Prime numbers. Next term: T (13).

Rule: The planets. Next term: J.

Rule: Directions. Next term: E.

Rule: The continents. Next term: E (Europe).

Task 6 Letter frequency

Answers will vary.

Task 7 Letter/Number codes

Fish

Elephant

School

Australia

Lynx

Zebra (alphabet backwards)

Cricket (alphabet backwards)

Kangaroo (alphabet backwards)

Kettle (alphabet backwards)

Maths (alphabet backwards)

You are very good at cracking codes!

Task 8 Letter shift codes

Balloon (1 letter back)

Birthday (1 letter back)

Computer (1 letter forward)

Puppy (1 letter back)

Help! (2 letters back)

Apple (2 letters forward)

Gorilla (3 letters forward)

France (3 letters forward)

Cubby (1 letter back) House (2 letters back)

Albert Einstein was great at cracking codes (1 letter back).

Task 9 Symbol codes

Eel

Ale

Last

Rust

Hand

"Hello" said the old sailor. "Can I stroll around the streets with you?"

"Sure" I said. "I live in this town."

"What is this town called?" asked the old sailor.

"It is called West Harbour" I said.

"Ah, there's buried treasure here" he said.

Task 10 Create a code

Answers will vary.

OXFORD UNIVERSITY PRESS

UNIT 3 – IT ALL ADDS UP
Addition

Addition is arguably the most useful of the four operations because it is the most used, and addition is also the basis for subtraction (addition in reverse), multiplication (repeated addition) and division (repeated subtraction). The commutative and associative laws of addition enable students to rearrange addition sums to create useful patterns and groupings. For example, it is easier to add 4 + 8 + 6 + 2 in a 'bonding to 10' approach of (4 + 6) + (8 + 2) and a statement like 6 + 6 + 5 + 6 might also be approached in the form of 3 × 6 + 5, or as 4 × 6 – 1.

Students should be encouraged to scan addition questions before doing them, so they might see 'short cuts' that involve bonding strategies, doubles or near doubles, or tables facts. The adage of 'keep it simple' (KIS) should always be emphasised. Brainstorming will reveal a myriad of examples where addition is used commonly in the community: from shopping, to counting steps or collections, to sports scores and tables/ladders.

TASK 1
What does it mean?
Understanding the concept of addition

There are lots of words and symbols in maths that mean 'to add'. Can you think of some?

Discuss examples of when and why we use addition. Alternatively, encourage students to draw a picture of something they might add up at home or at school.

TASK 2
Stepping out
Estimating and counting steps
You will need:

* BLM 10

Remind students that counting in ones is also the same as adding on one number at a time. Counting steps is a good way of practising counting in ones. Encourage the students to use their normal steps and not giant or mini steps.

Ask students to estimate and measure distances around school and home, for example:

1. How many steps from your front door to your back door?
2. How many steps from your bedroom to your kitchen?
3. How many steps from your letter box to your front door?

4. How many steps from your back door to your back fence?
5. How many steps across your back yard?
6. How many steps from your letterbox to your next-door neighbour's letter box?
7. How many steps from your front door to your next-door neighbour's letter box? (Look for a short cut.)
8. How many steps from your letter box to the end of your street?

TASK 3

Roll to 100

Adding single digits to 100

You will need:

- dice
- counters
- BLM 11

Students can play this game in pairs or by themselves.

Place your counters on the start position (1). If you are playing as a pair, use 'rock, paper, scissors' to decide who has the first turn. The first player rolls the die. Try to work out which number you will land before you move your counter, by adding the number on the die to the number you are on. The next player has a turn. The first player to reach 100 wins.

100	99	98	97	96	95	94	93	92	91
81	82	83	84	85	86	87	88	89	90
80	79	78	77	76	75	74	73	72	71
61	62	63	64	65	66	67	68	69	70
60	59	58	57	56	55	54	53	52	51
41	42	43	44	45	46	47	48	49	50
40	39	38	37	36	35	34	33	32	31
21	22	23	24	25	26	27	28	29	30
20	19	18	17	16	15	14	13	12	11
1 Start	2	3	4	5	6	7	8	9	10

What would be the fewest number of rolls you would need to make to reach 100 and win the game?

TASK 4

Let's play

Playing addition card games

You will need:

- deck of cards

Give pairs of students a deck of cards. Remove the picture cards (jacks, queens, kings, jokers). The aces will count as 1.

Make 10

This game is like 'Snap' but the aim is to make 10.

Remove the four 10 cards then shuffle the remaining cards and deal them out equally.

Take turns to place a card, face up, onto a pile.

Whenever two or more cards in a row add up to 10, be the first to say '10!' (For example, a 2 then an 8 add up to 10, and a 2 then a 3 then a 5 add up to 10, and a 2, an ace, a 3 and a 4 add up to 10).

If you are correct, you keep the cards in the pile. If you are wrong, the other player gets to keep the cards in the pile.

The winner is the player who gets all the cards in the pile and there are no more cards left.

Two then one

In this game, you add a single-digit number to a 2-digit number.

Remove the four 10 cards.

Shuffle the remaining cards and deal them out equally.

Player 1 puts down two cards making a 2-digit number (e.g. 3 then 6 makes 36).

Player 2 puts down another card.

The first player to find the total takes all three cards.

For the next round, the Player 2 puts down two cards and Player 1 puts down another card.

The winner is the player who wins all cards and there are no more cards left.

Stop!

In this game, you use addition to get as close to a total as you can.

Agree on a target between 20 and 100.

Player 1 shuffles the cards then turns them over one at a time, while Player 2 adds them up.

When the total gets close to the target, Player 2 says 'Stop!'.

Then swap so Player 2 turns the cards over and Player 1 adds them up.

The winner is the player who calls 'Stop!' when the total is on or closest to the target.

For the next round, choose a different target.

Bad card

In this game, you use addition and chance.

Player 1 shuffles the cards and places them face down.

Player 2 chooses a 'bad card' somewhere between 1 and 10.

Player 1 then turns over all the cards, while Player 2 adds them up.

The round ends when the 'bad card' is turned over.

Then the players swap roles and repeat.

The player with the highest total when the 'bad card' appears is the winner.

TASK 5

Short cuts

Using addition tricks and tips

You will need:

• deck of cards

Explain to the students that whenever we add up, we should begin by looking at the numbers we are dealing with.

You might find a short cut by seeing a pattern or using your addition skills and strategies. This way you 'keep it simple' (KIS). The simpler the problem, the more likely it is that you will get it right!

Bonding to 10 or Friends of 10

Use a set of cards from the ace (1) to 9.

Shuffle the cards and lay them face up.

Now pair up the cards that bond to 10.

What is the total of the 9 cards?

Consecutive addition

Place the cards in a row from ace (1) to 9.

Numbers that follow on from each other are called consecutive numbers.

The easiest way of adding up a group of consecutive numbers is to find the middle number and multiply by how many numbers there are.

So, in 1 to 9, the middle number is 5, and $9 \times 5 = 45$.

Tables addition

9 × 5 really means 5 + 5 + 5 + 5 + 5 + 5 + 5 + 5 + 5.

If you have to add up numbers that are close to tables, like 6 + 7 + 6 + 6, the easiest way to add them up is to use tables facts, so 4 × 6 + 1, or 24 + 1.

Now work in pairs.

Ask your partner to say five different numbers. (First, decide whether they will be 1-digit, 2-digit or even 3-digit numbers.)

Write the numbers down. Use patterns and the strategies discussed above to find the sum of these five numbers.

What strategies did you use to find the answer?

Now swap roles and repeat the challenge.

TASK 6

Nine-card cross

Solving addition puzzles

You will need:

- deck of cards

Show students a set of cards numbered 1 (ace) to 9. Lay the cards out to form a cross that looks like this:

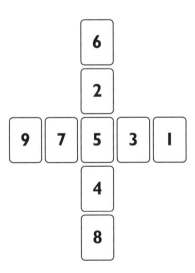

Add up the horizontal numbers with the students (25). Now add up the vertical numbers with the students (25). Explain that because the horizontal and vertical totals are the same, this is called a 9-card cross.

Use a set of cards numbered 1 (ace) to 9 to make 9-card crosses that have a 1, 3, 7 or 9 in the middle.

TASK 7
Pay the bill
Rounding to find totals

Sometimes we don't need to find an exact answer to an addition problem. We can round numbers up or down to get a rough idea of what a total might be.

1. Gemma is making fruit salad for a large party and wants to know about how much it will cost. Help Gemma out by rounding each item to the nearest dollar and estimating the total.

Pineapple	$3.40
Apples	$2.75
Grapes	$1.65
Kiwi fruit	$4.25
Apricots	$3.90
Melon	$5.60
Passionfruit	$6.15
Strawberries	$5.95

Now use your addition skills to find the exact cost of the fruits. What is the difference?

2. Sam is making vegetable soup for the same party and wants to know about how much it will cost. Help Sam out by rounding each item to the nearest dollar and estimating the total.

Potatoes	$3.80
Pumpkin	$2.55
Zucchini	$1.85
Carrots	$2.25
Corn	$3.75
Beans	$6.80
Onions	$4.15
Garlic	$7.95
Peas	$4.75
Broccoli	$4.45

Now use your addition skills to find the exact cost of the vegetables. What is the difference?

TASK 8

Prime numbers and perfect squares

Adding perfect squares to make a prime number

You will need:

*   BLM 12

Pierre de Fermat was a French mathematician who discovered that many prime numbers can be shown to be made up as the sum of two different perfect squares. Remind the students that a prime number is one that can only be divided by 1 and itself (so 1, 2, 3, 5, 7, 11 and so on) and a perfect square is a number multiplied by itself (so 2 × 2, 3 × 3, 4 × 4 and so on). So, the prime number 5 can be made up as the sum of two perfect squares: 1 × 1 + 2 × 2.

How many of these prime numbers can be made up as the sum of two perfect squares?

Number	Yes or No?	How?
5	Yes	1 × 1 + 2 × 2
7		
11		
13		
17		
19		
23		
29		
31		
37		
41		
43		
47		
53		
59		
61		
67		
71		
73		
79		

Number	Yes or No?	How?
83		
89		
97		

TASK 9

Problems to solve

Applying addition skills to unfamiliar situations

Encourage students to try these challenging addition problems:

1. The sum of four consecutive numbers is 150. What is the biggest of these four numbers?

2. Add up all the numbers from 1 to 99. What strategy did you use?

3. Averages are found by adding a group of numbers together and dividing that total by the amount of numbers in the group.

   After four innings, a cricketer has a batting average of 60. When they complete their next innings their average drops to 55.

   How many runs did the cricketer score in their 5th innings?

4. In maths, a palindromic number is the same if you read it forwards and backwards (e.g. 77, 191 or 4554).

   You can create a palindromic number by adding one number and the same number backwards (e.g. 34 + 43 = 77).

   Sometimes you need to keep adding like this several times until you get to the palindromic number. For example, 57+75 = 132, but 132 is not a palindromic number, so keep adding: 132 + 231 = 363, and that is a palindromic number!

   Start with the number 180 to create a palindromic number.

5. In this addition sum, different letters represent different digits. If a letter is repeated, a digit will be repeated.

   What are the two possible values for EVEN?

   ```
     O D D
   + O D D
   ─────────
   E V E N
   ```

ANSWERS

Task 1 What does it mean?

Answers will vary but terms to add could include 'sum', 'total', 'altogether', 'plus', 'and', 'collect', 'gather'.

Answers will vary but may include scoring in games, totals in sport, shopping bills, steps, etc.

OXFORD UNIVERSITY PRESS

Task 2 Stepping out

Answers will vary.

Task 3 Roll to 100

The fewest number of rolls needed to get to 100 is 17. This can be done with 16 rolls of 6 and a roll of 4.

Task 4 Let's play

Answers will vary.

Task 5 Short cuts

Bonding to 10 or Friends of 10: 45 (1 + 2 + 3 + 4 + 5 + 6 + 7 + 8 + 9)
Other answers will vary.

Task 6 Nine-card cross

There are multiple answers such as:

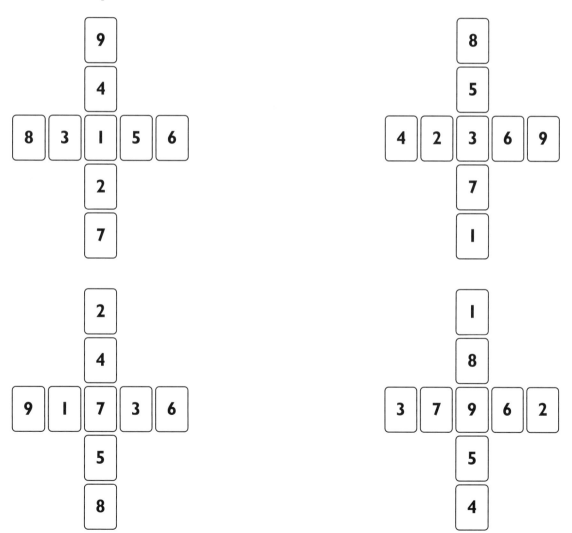

Task 7 Pay the bill

1. The rounded total is $34. The real total is $33.65. The difference is $0.35.
2. The rounded total is $43. The real total is $42.30. The difference is $0.70.

Task 8 Prime numbers and perfect squares

Number	Yes or No?	Proof if Yes
5	Yes	1 + 4
7	No	
11	No	
13	Yes	4 + 9
17	Yes	1 +16
19	No	
23	No	
29	Yes	4 + 25
31	No	
37	Yes	1 + 36
41	Yes	16 + 25
43	No	
47	No	
53	Yes	4 + 49
59	No	
61	Yes	25 + 36
67	No	
71	No	
73	Yes	9 + 64
79	No	
83	No	
89	Yes	25 + 64
97	Yes	16 + 81

OXFORD UNIVERSITY PRESS

Task 9 Problems to solve

1. The midpoint of the 4 numbers must be $150 \div 4$ or 37.5. Thus the 4 numbers must be 36, 37, 38 and 39 with the biggest being 39.

2. Because1 to 99 represents consecutive numbers, the midpoint must be 50. $99 \times 50 = 4950$.

3. If the cricketer's average is 60 after 4 innings, they must have made 240 runs in total because $240 \div 4 = 60$. After 5 innings their total number of runs must be 55×5 or 275. The cricketer must have scored 35 runs in their 5th innings.

4. $180 + 081 = 261$
   $261 + 162 = 423$
   $423 + 324 = 747$ (a palindromic number).

5. E must equal 1, because two three-digit numbers added together cannot equal more than 1998. D must equal 5, because in the tens place together they equal 1 as well, thanks to the carry over from the ones place. This makes N equal 0. This means O, which has to be bigger than 5, can equal 6 or 8. Thus, EVEN could equal 1310 or 1710.

UNIT 4 – TAKE IT AWAY
Subtraction

Subtraction is the inverse operation of addition. Anyone who can add, can subtract. There are many ways of finding the difference between two numbers, but the most common way is by addition. If you go into a shop with a $10 note and buy a toy for $8, your change must be $2 because your change and the price of the toy you bought must add up to the money you started with (10 – 8 = 2 because 2 + 8 = 10).

In fact, most subtraction we do in our everyday lives, we do as addition. Encourage your students to be as flexible with their approach to subtraction as they are with their approach to addition. The better they become at mental arithmetic, the more likely they will do subtraction questions as addition problems. If the numbers we are dealing with are not close together, then an algorithm, such as 'decomposition' or 'renaming', may need to be used, drawing upon a student's knowledge of place value. For example, 623 – 476, with 623 eventually 'decomposing' to 500 + 110 + 13 (same total but different place value structure).

Practise basic subtraction facts with your students, looking for patterns at every opportunity, such as: –9 = – 10 + 1. During brainstorming sessions, encourage your students to consider how subtraction is vital: when shopping to calculate change, when finding differences in sporting totals or times, when calculating how many marks are lost on tests, or when calculating ages based on birth dates.

Of course, we do not always subtract from whole numbers. For example, when we calculate change while shopping, we often subtract decimals. When we need to work out how much time is left in a game of netball or football, we might subtract fractions from whole numbers.

The tasks in this unit deal with finding differences applied to rational numbers (fractions and decimals). And although students do not really study negative numbers in depth until secondary school, they may have a working knowledge of negative numbers from the world around us – subzero temperatures, for example. If students get the idea that 6 – 7 = –1, they might be ready to see and work with the most amazing way to subtract whole numbers – subtracting by negatives.

TASK 1

What does it mean?

Understanding the concept of subtraction

1. There are lots of words and symbols in maths that mean 'to subtract'. Can you think of some?
2. Can you think of examples of when and why we use subtraction?

3. Use these prompts to make a number story about subtraction:
   (a) $10, newsagent, magazine $6, $4 change
   (b) Swifts, 65 goals, Firebirds 73 goals, 8 goals
   (c) 2019, Cat Foods Are Us, $28 000; 2020 $19 000

TASK 2

Fun facts

Subtracting from 10, 100 and 1000

You will need:

- BLM 13

Ask students to fill in these difference tables.

1. Start at the number in the first column and then count up to the bigger number in the top row to find the difference. You can use a ruler to help you. The first row has been done for you.

Difference	10	7	12	8	11	9	6	20
5	5	2	7	3	6	4	1	15
2								
1								
3								
4								
0								
6								

2. Finding the difference between a number and 100 is a common practice. Can you subtract these numbers from 100? The first row has been done for you.

From 100

Subtract	Answer	Subtract	Answer	Subtract	Answer	Subtract	Answer
95	5	52	48	24	76	10	90
67		49		75		65	
41		82		91		33	
79		22		47		77	
15		98		54		5	
85		2		50		40	
21		35		61		88	
58		70		16		20	
25		8		81		63	
90		45		39		18	

3. Can you subtract these numbers from 1000? The first row has been done for you.

From 1000

Subtract	Answer	Subtract	Answer	Subtract	Answer	Subtract	Answer	Subtract	Answer
998	2	110	890	40	960	655	345	730	270
402		88		596		201		698	
715		950		98		888		440	
4		333		797		190		155	
215		620		296		375		830	
550		495		911		510		399	

TASK 3

Backwards addition

Connecting addition and subtraction

You will need:

- cards numbered 1 to 20
- BLM 14

Give students a set of number cards from 1 to 20.

1. Shuffle the cards and select two cards at random. Now find the difference between the larger number and the smaller number by adding on from the smaller number until you reach the larger number. Record your results in the table.

Larger number	Smaller number	Subtraction question	Addition question	Answer
7	5	7 – 5	5 + 7	12

Make the task harder for the students by using 2-digit numbers.

OXFORD UNIVERSITY PRESS

2. Take the cards numbered 1 to 9 and choose four of them. Using the four cards, make two 2-digit numbers. Now find the difference between the larger number and the smaller number by adding on from the smaller number until you reach the larger number. Record your results in the table.

Larger number	Smaller number	Subtraction question	Addition question	Answer
19	12	19 – 12	12 + 19	31

TASK 4

Let's play

Playing subtraction games

You will need:

- playing cards
- calculator
- cards numbered 1 to 20
- dice

Subtraction snap

In this game ace is 1, 2 to 10 are face value, jack is 11, queen 12 and king 13. Shuffle the cards and deal them out equally.

Put your cards face down and take turns to lay a card face up.

The first player to say the difference between the 2 cards keeps both.

The winner of the game has the most cards after the deck has been worked through.

Take it away

Play with a partner.

Type 21 onto a calculator.

Take turns to take away 1, 2 or 3 from the number. The first player to reach zero wins the game.

(After you get good at this game, change both the starting number and what you can take away; for example, start at 46 and take away 1, 2, 3, 4 and 5.)

100 down

Play with a partner.

Use cards numbered 1 to 20.

Start at 100.

Take turns to select a card and take that number off progressively from what is left until zero is reached or a number below zero is reached.

Zero or bust

Play with a partner. Get a 6-sided die.

In this game you each start with 100 points and have 7 rolls to try, by using subtraction, to make zero or as close to zero as you can.

Each time you roll the dice you can take the number rolled as itself or 10 times itself. (A roll of 6 can be seen as 6 or 60.)

Keep score on a piece of paper.

If you go below zero you immediately lose the game.

10 different

Get cards numbered 1 to 20.

Lay the cards, face up on a table.

Your task is to select 10 different cards, each time removing the number on the card from 100. Your aim is to get as close as possible to zero.

When you have achieved this by reaching zero, aim to do the same with 9 cards. Then aim to do the same with 8 cards.

What are the fewest number of cards that need to be chosen to reach zero?

OXFORD UNIVERSITY PRESS

TASK 5

On sale

Working out savings with subtraction

You will need:

- BLM 15

Explain to students that a shop is having a sale. They can use their subtraction skills to work out the answers to these questions, using the table to help organise their thinking.

1. Soft toys

At the Kidz Toys store, soft toys cost $8 each. You buy one toy at full price. You buy five toys on sale at half-price. What is the total cost? How much did you save?

Product	Normal price	Sale price	Subtraction	Saving
Soft toy	$8	$4	8 – 4	$4

2. Toy trucks

Toy trucks cost $12 each. On sale, every truck is half price. You buy 12. What is the total cost? How much did you save?

3. Dolls

Dolls cost $11 each. The sale price is: $1 off the first doll bought; $2 off the second doll bought; $3 off the third doll bought; every subsequent doll bought is half normal price. You buy 7. What is the total cost? How much did you save?

4. Supermarket specials catalogue

Find a supermarket or shop catalogue, or search for a catalogue online. Use the information found in the catalogue and your subtraction skills to complete the table on BLM 15.

TASK 6

The number for one

Subtracting fractions and decimals

You will need:

- BLM 16

Explain to students that when we subtract a fraction from a whole number, we need to consider the appropriate fractional name for the whole number that we are dealing with. For example, the number 1 can be named as many different fractions: $\frac{2}{2}, \frac{3}{3}, \frac{4}{4}$ and so on.

Complete this table by finding the name for the number 1, then doing the subtraction. The first one has been done for you.

Question	Name for 1	Answer	Question	Name for 1	Answer
$1-\dfrac{3}{4}$	$\dfrac{4}{4}$	$\dfrac{1}{4}$	$1-\dfrac{1}{3}$		
$1-\dfrac{2}{5}$			$1-\dfrac{7}{10}$		
$1-\dfrac{1}{6}$			$1-\dfrac{5}{8}$		
$1-\dfrac{4}{9}$			$1-\dfrac{5}{12}$		

Complete this table by using the name for the number 1 to work out the answers in these subtraction problems with mixed numbers. The first one has been done for you.

Question	Name for 1	Answer	Question	Name for 1	Answer
$2-\dfrac{7}{8}$	$\dfrac{8}{8}$	$1\dfrac{1}{8}$	$2-\dfrac{2}{3}$		
$3-\dfrac{1}{4}$			$3-\dfrac{4}{5}$		
$2-\dfrac{5}{6}$			$4-\dfrac{5}{7}$		
$10-\dfrac{7}{12}$			$8-\dfrac{4}{11}$		

Use your knowledge of how decimals work to find the answers to these subtraction problems. The first one has been done for you. Remember again that you only ever need to change 1 whole number when subtracting from a whole number and decimal. For instance, for the question $8.5 - 4.8$, because 1 whole number equals $\dfrac{10}{10}$, 8.5 will rename to 7 and $\dfrac{15}{10}$. So, 7 and $\dfrac{15}{10}$ – 4 and $\dfrac{8}{10}$ must equal 3 and $\dfrac{7}{10}$ or 3.7. The first one has been done for you.

Question	Name for 1	Answer	Question	Name for 1	Answer
1 – 0.6	$\dfrac{10}{10}$	0.4	1 – 0.3		
2 – 0.7			2 – 0.8		
6 – 1.4			5 – 1.2		
10 – 2.3			9 – 4.5		
1 – 0.04			1 – 0.23		
2 – 0.06			5 – 0.01		
3 – 1.12			7 – 5.55		

OXFORD UNIVERSITY PRESS

TASK 7

On target

Skip counting backwards

Skip counting is something we often do, counting forwards to help with addition: 3, 6, 9, 12, 15 … But what if we skip count backwards? In this task, students will need to use their knowledge of subtraction, their common sense and their ability to see a pattern to fill in the gaps.

The target is 23. Will you hit 23 in these backwards skip counts?

1. 99, 97, 95, 93 …
2. 99, 96, 93, 90 …
3. 100, 96, 92, 88 …
4. 98, 93, 88, 83 …
5. 100, 93, 86, 79 …
6. 71, 65, 59, 53 …
7. 123, 112, 101, 90 …
8. 1123, 1113, 1103, 1093 …

9. 102, 94, 86, 78 …
10. 104, 95, 86, 77 …
11. 143, 131, 119, 107 …
12. 150, 135, 120, 105 …
13. 166, 153, 140, 127 …
14. 230, 207, 184, 161 …
15. 233, 218, 203, 188 …

TASK 8

Subtraction strategies

Topping up and using equal differences

You will need:

* BLM 17

There are many ways of doing subtraction. Encourage the students to try out a variety of different strategies.

'Topping up' is finding the difference between two numbers by using addition. Think about filling up a container that already has some water in it, right up to the top!

> *Example of how it works:*
> *102 – 87:*
> *Step 1. 87 to 100 = 13*
> *Step 2. 100 to 102 = 2*
> *Step 3. Answer: 13 + 2 = 15.*

1. Can you complete the table on BLM 17 by copying the steps above?

Question	Step 1	Step 2	Step 3 Answer
105 – 97			
210 – 186			
311 – 88			
407 – 299			

Question	Step 1	Step 2	Step 3 Answer
514 – 478			
616 – 497			
1123 – 998			

Using the 'equal differences' strategy can help students with questions that would normally involve renaming, which involves lots of crossing outs and exchanges, like 1002 – 674.

It's a mathematical fact that the difference between numbers will always remain the same if you subtract both numbers by the same amount. So, for example, the difference between 10 and 8 and 6 and 4 (2) will be the same because the amount removed from both numbers (4) was the same.

> *Example of how it works:*
> *1002 – 674*
> *Step 1: Try to make 1002 into a number that won't need renaming. If you take away 3 it will become 999.*
> *Step 2: Take 3 from 674 making 671.*
> *Step 3 Answer: 999 – 671 = 328 (same answer as 1002 – 674)*

2. Can you complete the table on BLM 17 by copying the steps above?

Question	Step 1	Step 2	Step 3 Answer
405 – 278			
701 – 567			
303 – 129			
817 – 778			
1022 – 893			
1006 – 947			
10 005 – 9846			

TASK 9

In the negative

Subtracting with negative numbers

You will need:

- BLM 18

Practice some simple 1-digit negative subtractions with the students, such as 5 – 6, 4 – 8, 3 – 5 and so on. If students can understand the idea that 6 – 7 = –1, challenge them to use negative numbers when working with 2- and 3-digit numbers.

> *Example of how it works:*
> *324 – 186*
> *Step 1: 300 – 100 = 200*
> *Step 2: 20 – 80 = –60*
> *Step 3: 4 – 6 = –2*
> *Step 4 (answer) 200 – 60 – 2 = 138.*

OXFORD UNIVERSITY PRESS

1. Follow this method to solve these 2-digit subtractions. The first one has been done for you.

Question	Step 1	Step 2	Answer
56 – 29	50 – 20 = 30	6 – 9 = –3	30 – 3 = 27
73 – 25			
83 – 56			
92 – 45			
53 – 44			

2. Try some 3-digit subtraction sums. The first one has been done for you.

Question	Step 1	Step 2	Step 3	Answer
213 – 138	200 – 100 = 100	10 – 30 = –20	3 – 8 = –5	100 – 20 – 5 = 75
156 – 88				
246 – 177				
452 – 289				
751 – 532				

3. Try some 4-digit subtraction sums. The first one has been done for you.

Question	Step 1	Step 2	Step 3	Step 4	Answer
3192 – 1476	3000 – 1000 = 2000	100 – 400 = –300	90 – 70 = 20	2 – 6 = –4	2000 – 300 + 20 – 4 = 1716
4561 – 3778					
5234 – 2881					
8688 – 2935					
7346 – 6882					

TASK 10

Problems to solve

Applying subtraction skills to unfamiliar situations

You will need:

- calculator
- BLM 19

Encourage students to try these challenging subtraction problems:

1. Frank starts at 0 and counts up by 5s. At the same time, Ernest starts at 100 and counts down by 3s. Will Frank and Ernest ever say the same number at the same time?

2. What number gives the same result when either 20 is subtracted from it or when it is divided by 6?

3. In this question, every letter stands for a different digit:

   B I K E − C A R = EM

   What is the smallest possible value of EM?

4. Penny wanted to subtract a single-digit number from a number in the 60s using her calculator but she hit the × button rather than the − button. The number on her screen was 537 larger that it should have been.

   What should Penny have typed onto her calculator?

5. In which column would the number 19 appear in this countdown table?

A	B	C	D	E	F	G	H	I
1000	995	998	996	994	999	992	993	997
991	986	989	987	985	990	983	984	988
982	977	980	978	976	981	974	975	979
973	968	971	969	967	972	965	966	970

ANSWERS

Task 1 What does it mean?

1. Take away, difference, reduce, remove, less than, minus, negative, count back, margin, etc.

2. When shopping and giving change, when finding out how many more runs a team needs to score in a game of cricket, goals to score in a game of netball, when finding out the margin between 2 teams, when calculating how much a temperature has dropped, when finding out how much more needs to be saved to buy something, when comparing 2 or more items for size or how full they are, etc.

3. Possible answers:

   (a) I went to the newsagent and bought a magazine costing $6. I paid with a $10 note and received $4 in change.

   (b) The Firebirds beat the Swifts 73 goals to 65 goals, with a winning margin of 8 goals.

   (c) In 2019 Cat Foods Are Us made $28 000 profit, but this dropped by $9000 in 2020 to $19 000.

Task 2 Fun facts

1.

Difference	10	7	12	8	11	9	6	20
5	5	2	7	3	6	4	1	15
2	8	5	10	6	9	7	4	18

OXFORD UNIVERSITY PRESS

Difference	10	7	12	8	11	9	6	20
1	9	6	11	7	10	8	5	19
3	7	4	9	5	8	6	3	17
4	6	3	8	4	7	5	2	16
0	10	7	12	8	11	9	6	20
6	4	1	6	2	5	3	0	14

2.

From 100

Subtract	Answer	Subtract	Answer	Subtract	Answer	Subtract	Answer
95	5	52	48	24	76	10	90
67	33	49	51	75	25	65	35
41	59	82	18	91	9	33	67
79	21	22	78	47	53	77	23
15	85	98	2	54	46	5	95
85	15	2	98	50	50	40	60
21	79	35	65	61	39	88	12
58	42	70	30	16	84	20	80
25	75	8	92	81	19	63	37
90	10	45	55	39	61	18	82

3.

From 1000

Subtract	Answer	Subtract	Answer	Subtract	Answer	Subtract	Answer	Subtract	Answer
998	2	110	890	40	960	655	345	730	270
402	598	88	912	596	404	201	799	698	302
715	285	950	50	98	902	888	112	440	560
4	996	333	667	797	203	190	810	155	845
215	785	620	380	296	704	375	625	830	170
550	450	495	505	911	89	510	490	399	601

Task 3 Backwards addition

1. Answers will vary.

2. Answers will vary.

Task 4 Let's play

In the first four games the answers will vary.

In the game '10 different', the fewest number of cards required to reach zero is 6. For example: $100 -20, -19, -18, -17, -16, -10$.

Task 5 On sale

1. $28 cost. $20 saved.

2. $72 cost. $72 saved.

3. $49 cost. $28 saved.

4. Answers will vary.

Task 6 The number for one

Question	Name for 1	Answer	Question	Name for 1	Answer
$1 - \frac{3}{4}$	$\frac{4}{4}$	$\frac{1}{4}$	$1 - \frac{1}{3}$	$\frac{3}{3}$	$\frac{2}{3}$
$1 - \frac{2}{5}$	$\frac{5}{5}$	$\frac{3}{5}$	$1 - \frac{7}{10}$	$\frac{10}{10}$	$\frac{3}{10}$
$1 - \frac{1}{6}$	$\frac{6}{6}$	$\frac{5}{6}$	$1 - \frac{5}{8}$	$\frac{8}{8}$	$\frac{3}{8}$
$1 - \frac{4}{9}$	$\frac{9}{9}$	$\frac{5}{9}$	$1 - \frac{5}{12}$	$\frac{12}{12}$	$\frac{7}{12}$

Question	Name for 1	Answer	Question	Name for 1	Answer
$2 - \frac{7}{8}$	$\frac{8}{8}$	$1\frac{1}{8}$	$2 - \frac{2}{3}$	$\frac{3}{3}$	$1\frac{1}{3}$
$3 - \frac{1}{4}$	$\frac{4}{4}$	$2\frac{3}{4}$	$3 - \frac{4}{5}$	$\frac{5}{5}$	$2\frac{1}{5}$
$2 - \frac{5}{6}$	$\frac{6}{6}$	$1\frac{1}{6}$	$4 - \frac{5}{7}$	$\frac{7}{7}$	$3\frac{2}{7}$
$10 - \frac{7}{12}$	$\frac{12}{12}$	$9\frac{5}{12}$	$8 - \frac{4}{11}$	$\frac{11}{11}$	$7\frac{7}{11}$

Question	Name for 1	Answer	Question	Name for 1	Answer
$1 - 0.6$	$\frac{10}{10}$	0.4	$1 - 0.3$	$\frac{10}{10}$	0.7
$2 - 0.7$	$\frac{10}{10}$	1.3	$2 - 0.8$	$\frac{10}{10}$	1.2
$6 - 1.4$	$\frac{10}{10}$	4.6	$5 - 1.2$	$\frac{10}{10}$	3.8
$10 - 2.3$	$\frac{10}{10}$	7.7	$9 - 4.5$	$\frac{10}{10}$	4.5
$1 - 0.04$	$\frac{100}{100}$	0.96	$1 - 0.23$	$\frac{100}{100}$	0.77

OXFORD UNIVERSITY PRESS

Question	Name for 1	Answer	Question	Name for 1	Answer
2 – 0.06	$\frac{100}{100}$	1.94	5 – 0.01	$\frac{100}{100}$	4.99
3 – 1.12	$\frac{100}{100}$	1.88	7 – 5.55	$\frac{100}{100}$	1.45

Task 7 On target

1. Yes
2. No
3. No
4. Yes
5. Yes
6. Yes
7. No
8. Yes
9. No
10. Yes
11. Yes
12. No
13. Yes
14. Yes
15. Yes

Task 8 Subtraction strategies

1.

Question	Step 1	Step 2	Step 3 Answer
105 – 97	97 to 100 = 3	100 to 105 = 5	3 + 5 = 8
210 – 186	186 to 200 = 14	200 to 210 = 10	14 + 10 = 24
311 – 88	88 to 300 = 212	300 to 311 = 11	212 + 11 = 223
407 – 299	299 to 400 = 101	400 to 407 = 7	101 + 7 = 108
514 – 478	478 to 500 = 22	500 to 514 = 14	22 + 14 = 36
616 – 497	497 to 600 = 103	600 to 616 = 16	103 + 16 = 119
1123 – 998	998 to 1100 = 102	1100 to 1123 = 23	102 + 23 =125

2.

Question	Step 1	Step 2	Step 3 Answer
405 – 278	Make 405 into 399 (–6)	Make 278 into 272 (–6)	399 –272 = 127
701 – 567	Make 701 into 699 (–2)	Make 567 into 565 (–2)	699 – 565 = 134
303 – 129	Make 303 into 299 (–4)	Make 129 into 125 (–4)	299 – 125 = 174
817 – 778	Make 817 into 799 (–18)	Make 778 into 760 (–18)	799 – 760 = 39
1022 – 893	Make 1022 into 999 (–23)	Make 893 into 870 (–23)	999 – 870 = 129
1006 – 947	Make 1006 into 999 (–7)	Make 947 into 940 (–7)	999 – 940 = 59
10 005 – 9846	Make 10 005 into 9 999 (–6)	Make 9846 into 9840 (–6)	9999 – 9840 = 159

Task 9 In the negative

1.

Question	Step 1	Step 2	Answer
56 – 29	50 – 20 = 30	6 – 9 = –3	30 – 3 = 27
73 – 25	70 – 20 = 50	3 – 5 = –2	50 –2 = 48
83 – 56	80 – 50 = 30	3 – 6 = –3	30 – 3 = 27
92 – 45	90 – 40 = 50	2 – 5 = –3	50 – 3 = 47
53 – 44	50 – 40 = 10	3 – 4 = –1	10 – 1 = 9

2.

Question	Step 1	Step 2	Step 3	Answer
213 – 138	200 – 100 = 100	10 – 30 = –20	3 – 8 = –5	100 – 20 – 5 = 75
156 – 88	100 – 0 = 100	50 – 80 = –30	6 – 8 = –2	100 – 30 – 2 = 68
246 – 177	200 – 100 = 100	40 – 70 = –30	6 – 7 = –1	100 – 30 – 1 = 69
452 – 289	400 – 200 = 200	50 – 80 = –30	2 – 9 = –7	200 – 30 – 7 = 163
751 – 532	700 – 500 = 200	50 – 30 = 20	1 – 2 = –1	200 + 20 – 1 = 219

3.

Question	Step 1	Step 2	Step 3	Step 4	Answer
3192 – 1476	3000 – 1000 = 2000	100 – 400 = –300	90 – 70 = 20	2 – 6 = –4	2000 – 300 + 20 – 4 = 1716
4561 – 3778	4000 – 3000 = 1000	500 – 700 = –200	60 – 70 = –10	1 – 8 = –7	1000 – 200 – 10 – 7 = 783
5234 – 2881	5000 – 2000 = 3000	200 – 800 = –600	30 – 80 = –50	4 – 1 = 3	3000 – 600 – 50 + 3 = 2353
8688 – 2935	8000 – 2000 = 6000	600 – 900 = –300	80 – 30 = 50	8 – 5 = 3	6000 – 300 + 50 + 3 = 5753
7346 – 6882	7000 – 6000 = 1000	300 – 800 = –500	40 – 80 = –40	6 – 2 = 4	1000 – 500 – 40 + 4 = 464

Task 10 Problems to solve

1. No, they will not ever say the same number at the same time. The closest they will get is 60 and 58.

2. 24 – 20 = 4 and 24 divided by 6 = 4.

3. 1023 – 987 would give the smallest possible difference of 36.

4. Penny should have typed 66 – 9 and not 66 × 9.

5. 19 is 1 more than a number in the 9 times table. Column A contains numbers that are 1 more than the 9 times table.

OXFORD UNIVERSITY PRESS

UNIT 5 – GOOD TIMES
Multiplication

The '×' sign means 'lots of' or 'groups of' and multiplication is merely repeated addition: 3 × 4 means 3 groups of 4, or 4 + 4 + 4, and is in the 4 times table. Thus, any laws of mathematics that apply to addition must likewise apply to multiplication, as they are identical operations. This fact enables students to be flexible and creative in their approach to multiplication, altering a problem to suit their own purposes.

For example, 17 × 5 (85) × 2 = 170 is so much more difficult if done inflexibly from left to right. Altering it to 5 × 2 (10) × 17 = 170 is so much easier. And 4 × 86 × 25 is a simple question if it is thought through logically … Try to make multiplication problems as easy as possible by looking for shortcuts and simplicity.

Anyone who can add up can quickly learn to multiply. This is true because multiplication is just addition done quickly. When multiplication was created, it was seen as a great way of making mathematics easier and faster.

When brainstorming with the students, demonstrate the fact that without multiplication it is very time consuming to find the total of a number of objects each costing a certain number of dollars or worth a certain number of points. The number of days in 6 weeks, 5c coins in a dollar, legs on 12 dogs, and months in 3 years can all be easily calculated with multiplication.

There is no substitute for knowing times tables – they are as important today as they ever were. When teaching your students the times tables facts, focus on the patterns contained within each one. Tests for divisibility show us, for example, that any number in the 3× table has its digits summing to 3, 6 or 9; any number in the 6× table will be even; and any number in the 9× table will have its digits summing to 9. You can also split up numbers to help when you multiply: 16 × 4 can be seen as 10 × 4 + 6 × 4 = 40 + 24 = 64. It is these patterns and connections that enable us to see how to get to an answer for many problems quickly and efficiently

Encourage rounding when teaching multiplication; it is something that we all do in the real world. Rounding can give us an acceptable and sensible answer. So 68 × 43 is about 70 × 40 and therefore about 2800. And 62 × 79 is about 6 tens × 8 tens. 6 × 8 = 48 and 10 × 10 = 100, so the answer is about 4800. The exact answer is 4898.

TASK 1
What does it mean?
Understanding the concept of multiplication

There are lots of words and symbols in maths that mean 'to multiply'. Can you think of some?

Discuss examples of when and why we use multiplication. Remind students that life (and mathematics) becomes easier as our knowledge of tables facts become automatic.

TASK 2

Repeated addition

Connecting multiplication and addition

Explain to students that the laws of addition (e.g. that the order does not matter when you add up numbers) can also apply to multiplication. So, if $8 + 12 = 12 + 8$, then 8×12 must equal 12×8. Remember that multiplication is addition, just done quickly.

1. Find a quick way of finding the answer to these questions:

   (a) $4 + 4 + 4$

   (b) $3 + 3 + 3 + 3$

   (c) $6 + 6 + 6 + 6 + 6$

   (d) $10 + 10 + 10 + 10 + 10 + 10 + 10$

   (e) $8 + 8 + 8 + 8 + 8 + 8$

   (f) $2 + 2 + 2 + 2 + 2 + 2 + 2 + 2 + 2 + 2 + 2$

2. Use your logical thinking and times tables skills to find easy ways of finding the answers to these questions:

   (a) $5 + 5 + 5 + 6 + 5$

   (b) $10 + 10 + 10 + 10 + 9 + 10$

   (c) $4 + 4 + 4 + 6 + 4 + 4.$

   (d) $3 + 3 + 3 + 5 + 5 + 5$

   (e) $6 + 6 + 7 + 6 + 6$

   (f) $9 + 9 + 9 + 9 + 9 + 7$

3. For these questions, see if you can find another way to find the answers – there may be more than one!

   (a) 13×4

   (b) 16×5

   (c) 17×9

   (d) 19×3

   (e) 22×8

   (f) 34×6

TASK 3

Tests for divisibility 1

Finding patterns in the 1, 2, 3, 5, 6, 9 and 10 times tables

You will need:

- BLM 20
- BLM 21

Remind students that when we count in groups, using times tables, patterns emerge.

Use a multiplication grid to help you find patterns in these times tables.

×	1	2	3	4	5	6	7	8	9	10
1	1	2	3	4	5	6	7	8	9	10
2	2	4	6	8	10	12	14	16	18	20
3	3	6	9	12	15	18	21	24	27	30
4	4	8	12	16	20	24	28	32	36	40
5	5	10	15	20	25	30	35	40	45	50
6	6	12	18	24	30	36	42	48	54	60
7	7	14	21	28	35	42	49	56	63	70
8	8	16	24	32	40	48	56	64	72	80
9	9	18	27	36	45	54	63	72	81	90
10	10	20	30	40	50	60	70	80	90	100

The 1 times table

How do you know if a number is in the 1 times table?

The 2 times table

How do you know if a number is in the 2 times table?

The 10 times table

How do you know if a number is in the 10 times table?

The 5 times table

How do you know if a number is in the 5 times table?

The 3 times table

How do you know if a number is in the 3 times table? (Clue: Add up the digits.)

The 6 times table

How do you know if a number is in the 6 times table? (Clue: See the 2× and 3× tables.)

The 9 times table

How do you know if a number is in the 9 times table? (Clue: Add up the digits.)

On BLM 21, use a tick to show when a number is in a times table and cross to show when it is not.

Number	1 times table	2 times table	3 times table	5 times table	6 times table	9 times table	10 times table
10							
21							
30							
36							
41							
54							
60							

Number	1 times table	2 times table	3 times table	5 times table	6 times table	9 times table	10 times table
77							
90							
102							
151							
366							

TASK 4

Tests for divisibility 2

Finding patterns in the 4, 7, 8, 11 and 12 times tables

You will need:

- BLM 20
- BLM 21
- calculator

Remind students that when we count in groups, using times tables, patterns emerge.

Use a multiplication grid to help you find patterns in these times tables.

×	1	2	3	4	5	6	7	8	9	10	11	12
1	1	2	3	4	5	6	7	8	9	10	11	12
2	2	4	6	8	10	12	14	16	18	20	22	24
3	3	6	9	12	15	18	21	24	27	30	33	36
4	4	8	12	16	20	24	28	32	36	40	44	48
5	5	10	15	20	25	30	35	40	45	50	55	60
6	6	12	18	24	30	36	42	48	54	60	66	72
7	7	14	21	28	35	42	49	56	63	70	77	84
8	8	16	24	32	40	48	56	64	72	80	88	96
9	9	18	27	36	45	54	63	72	81	90	99	108
10	10	20	30	40	50	60	70	80	90	100	110	120
11	11	22	33	44	55	66	77	88	99	110	121	132
12	12	24	36	48	60	72	84	96	108	120	132	144

The 4 times table

How do you know if a number is in the 4 times table? (Clue: The 2 times table.)

The 8 times table

How do you know if a number is in the 8 times table? (Clue: The 4 times table.)

The 12 times table

How do you know if a number is in the 12 times table? (Clue: The 3 and 4 times tables.)

OXFORD UNIVERSITY PRESS

The 11 times table

How do you know if a number is in the 11 times table? (Clue: The place of the digits.)

The 7 times table

How do you know if a number is in the 7 times table? (Clue: Double the ones digit.)

On BLM 21, use a tick to show when a number is in a times table and a cross to show when it is not.

Number	1 times table	2 times table	3 times table	4 times table	5 times table	6 times table	7 times table	8 times table	9 times table	10 times table	11 times table	12 times table
12												
29												
36												
54												
60												
73												
81												
100												
210												

*Can you find a number that is in **all** the times tables from 1 to 12? Use a calculator to help find an answer. Do you think this will be the first number in all the 12 tables?*

TASK 5

Sports scores

Multiplying to find totals in sport

You will need:

* BLM 22

Australian rules football

In AFL football, a goal is worth 6 points and a behind is worth 1 point (a 'behind' is when a player scores between a central and outer post). Which team won this game?

Team	Goals	Behinds	Points
Geelong	10	11	
West Coast Eagles	11	4	

Cricket

In cricket, batters can score 6s, 4s, 3s, 2s and 1s (singles). Which team won this game?

Team	6s	4s	3s	2s	1s	Total
South Africa	3	21	2	10	32	
India	1	30	1	8	42	

Rugby Union

In Rugby Union, a try is worth 5 points, a converted try is worth 2 points and a penalty is worth 3 points. Which team won this game?

Team	Tries	Conversions	Penalties	Score
England	6	4	3	
France	5	5	4	

Basketball

In basketball, a shot from behind the 3-point line is worth 3 points, a 'normal' shot is worth 2 points and a foul shot is worth 1 point. Which team won this game?

Team	3 pointers	Normal shot	Foul shots	Score
Chicago Bulls	8	33	6	
Detroit Pistons	6	38	7	

Gaelic football

In Irish or Gaelic football, a goal is worth 3 points and a point is worth 1 point. Which team won this game?

Team	Goals	Points	Total
Clontarf	5	4	
Cork	3	11	

TASK 6

Multiplying big numbers

Rounding to find sensible answers

You will need:

- BLM 23
- calculator

1. Use rounding to find answers to the problems on BLM 22 and then use a calculator to find the exact answer. The first one has been done for you.

Question	Rounded question	Rounded answer	Exact answer
54 × 67	50 × 70	3 500	3 618
21 × 67			
55 × 49			
378 × 4			
6 × 923			
109 × 43			
112 × 123			
226 × 589			
2102 × 6			
7989 × 21			
1076 × 268			

Why do you think the last rounded answer is a bit further away from the exact answer compared to the other questions?

2. The cost of entry at the show was $20.40 for adults and $4.90 for children. Mr Smith, Mrs Smith, Mr Smith's parents and the three Smith children all went to the show. Use rounding to find about how much it cost the Smith family to go to the show. Now use a calculator to find exactly how much it cost the Smiths to go to the show.

3. In 2020 there were 12 166 191 Australians in work. The average wage for these people was $78 954 per year. Use your rounding skills to find the total amount Australians earned in 2020. Can you find the exact answer using a calculator?

TASK 7

Short cuts

Using multiplication tricks and tips

Remind students that the simpler a problem is, the more likely we can find the correct answer.

Multiplying by 2

When we multiply a number by 2, we are simply doubling it. Multiply the following numbers by 2 by doubling each digit:

13

24

53

106

234

1124

2507

Multiplying by 4

When we multiply a number by 4, we double it and then double it again. Use this short cut to multiply these numbers by 4:

23

54

72

112

345

1345

2235

Multiplying by 5

Without doubt one of the easiest tables to learn is the 10 times table. Multiplying whole numbers by 10 simply asks us to add a zero. Often the easiest way to multiply by 5 is to multiply by 10 and then halve the answer. Use this strategy to multiply these numbers by 5:

14

26

45

62

123

344

1456

Multiplying by 9

Multiplying by 9 is as simple as multiplying by 10 and removing 1 group of the number you are dealing with. Use this strategy to multiply these numbers by 9:

13

24

37

58

120

230

561

Multiplying by 11

Like when we multiply by 9, multiplying by 11 uses the 'multiplying by 10' strategy. But this time, after multiplying the given number by 10, we add on the number we started with. Use this strategy to multiply these numbers by 11.

14

21

36

47

68

103

241

OXFORD UNIVERSITY PRESS

Left to right multiplication

Often it is easier to multiply left to right rather than right to left as we are used to doing. For example, 26 × 3 can be done as 20 × 3 + 6 × 3, equalling 78. Try the left to right strategy to find the answers to these questions.

32 × 4 = (_____ × 4) + (_____ × 4) =
53 × 7 = (_____ × 7) + (_____ × 7) =
59 × 12 = (_____ × 12) + (_____ × 12) =
42 × 6 = (_____ × 6) + (_____ × 6) =
66 × 3 = (_____ × 3) + (_____ × 3) =
82 × 11 = (_____ × 11) + (_____ × 11) =

TASK 8

Common multiples

Manipulating times tables facts

You will need:

• calculator

Remind students that once we know our times tables well, we can do some very impressive mathematics. One example of this is finding multiples and common multiples. A multiple of a number is its position in a times table – the 3rd multiple of 6 is 18. A common multiple is when numbers can be found in more than 1 table at the same time – 6 is in both the 2 times and the 3 times table and is called a common multiple of both 2 and 3.

1. List the 4th, 5th and 10th multiples of the following numbers:

5

11

3

12

8

20

2. List the 3rd, 9th and 30th multiples of the following numbers:

2

4

6

7

9

10

3. Can you find the first three common multiples of the following numbers:

2 and 3

4 and 6

5 and 7

2 and 3 and 4

2 and 4 and 5

4. Three lighthouses are on the surf coast and all flash at different times. Cape Easterly flashes every 3 minutes. Cape Central flashes every 4 minutes. Cape Westerly flashes every 5 minutes. If they all flash together at 12:30pm, at what time will they all flash together again?

TASK 9
The Sieve of Eratosthenes
Using times tables to find prime numbers
You will need:

- BLM 24

Explain to students that Eratosthenes (276–194 BC) was a Greek mathematician who created a way to find prime numbers by using multiplication tables. A prime number has only two factors, 1 and itself, and so is only in the 1 times table and the times table of itself. In this task, students will use what came to be known as 'The Sieve of Eratosthenes' to find all the prime numbers to 150 by using their times tables knowledge.

Steps

1. Cross out 1 because it only has 1 factor.
2. Leave 2 then cross out all other even numbers (this will also remove all numbers in the 4, 6, 8, 10 and 12 times tables).
3. Leave 3, 5, 7 and 11 and then cross out any numbers left in the 3, 5, 7 and 11 times tables.
4. What are left are prime numbers.

1	2	3	4	5	6	7	8	9	10
11	12	13	14	15	16	17	18	19	20
21	22	23	24	25	26	27	28	29	30
31	32	33	34	35	36	37	38	39	40
41	42	43	44	45	46	47	48	49	50
51	52	53	54	55	56	57	58	59	60
61	62	63	64	65	66	67	68	69	70
71	72	73	74	75	76	77	78	79	80
81	82	83	84	85	86	87	88	89	90
91	92	93	94	95	96	97	98	99	100
101	102	103	104	105	106	107	108	109	110
111	112	113	114	115	116	117	118	119	120
121	122	123	124	125	126	127	128	129	130
131	132	133	134	135	136	137	138	139	140
141	142	143	144	145	146	147	148	149	150

Write down all the prime numbers from 1 to 150. How many are there? How many are not prime numbers?

OXFORD UNIVERSITY PRESS

TASK 10
Problems to solve
Applying multiplication skills to unfamiliar situations

1. Sophie changes her books at the library every 4 days. Jack changes his books at the library every 5 days and Charlotte changes her books at the library every 6 days. If they were all at the library on 1 June, when will they all be at the library together again?

2. Sam's collection of Pokémon cards can be divided into:

   2 equal groups with 1 card left over

   3 equal groups with 1 card left over

   5 equal groups with 1 card left over

   7 equal groups with 1 card left over.

   What is the fewest number of Pokémon cards that Sam could own?

3. I have 24 pieces of timber, each 1 m long. I want to build a rectangular garden bed but do not want to cut any pieces of timber. What is the biggest area my garden bed could be?

4. Follow these rules:

   (a) Use the digits 1, 2, 3 and 4, NOT 2, 3, 4 and 5.

   (b) Make either a 3-digit by 1-digit or a 2-digit by 2-digit multiplication problem.

   (c) Which combination of digits will give you the biggest possible product?

5. In Australian rules football, a goal is worth 6 points and a behind is worth 1 point. So, for example 10 goals and 10 behinds would equal 70 points.

   2 goals and 12 behinds is called a 'tables' score because 2 goals and 12 behinds equal 24 points, and $2 \times 12 = 24$.

   There are only 3 other possible tables scores in Australian rules football. Can you find them?

ANSWERS

Task 1 What does it mean?

Answers will vary but may include:

<u>What?</u> Multiplication means: to find groups of things, to find collections of things, to add up equal groups quickly. Words that mean 'to multiply' include 'times', 'product', 'groups of', 'lots of'.

<u>When?</u> Finding totals in football, cricket, basketball, golf, rugby, finding crowd sizes, rounding, calculating bills, counting money, finding areas.

<u>Why?</u> We use multiplication to add up groups of things quickly; we use multiplication to make maths easier.

Task 2 Repeated addition

1. **(a)** $3 \times 4 = 12$.
   (b) $4 \times 3 = 12$.
   (c) $5 \times 6 = 30$.
   (d) $7 \times 10 = 70$.
   (e) $6 \times 8 = 48$.
   (f) $11 \times 2 = 22$.
2. **(a)** $5 \times 5 + 1 = 26$.
   (b) $6 \times 10 - 1 = 59$.
   (c) $6 \times 4 + 2 = 26$.
   (d) $3 \times 3 + 3 \times 5 = 24$.
   (e) $5 \times 6 + 1 = 31$.
   (f) $6 \times 9 - 2 = 52$.
3. **(a)** $10 \times 4 + 3 \times 4 = 52$.
   (b) $10 \times 5 + 6 \times 5 = 80$.
   (c) $10 \times 9 + 7 \times 9 = 153$.
   (d) $10 \times 3 + 9 \times 3 = 57$.
   (e) $10 \times 8 + 10 \times 8 + 2 \times 8 = 176$.
   (f) $10 \times 6 + 10 \times 6 + 10 \times 6 + 4 \times 6 = 204$.

Task 3 Tests for divisibility 1

The 1 times table: Every whole number is in the 1 times table.

The 2 times table: Every even number is in the 2 times table.

The 10 times table: Numbers ending in 0 are in the 10 times table.

The 5 times table: Numbers ending in 5 or 0 are in the 5 times table.

The 3 times table: Numbers whose digits sum to 3, 6 or 9 are in the 3 times table.

The 6 times table: Even numbers whose digits sum to 3, 6 or 9 are in the 6 times table.

The 9 times table: Numbers whose digits sum to 9 are in the 9 times table.

Number	1 times table	2 times table	3 times table	5 times table	6 times table	9 times table	10 times table
10	Yes	Yes	No	Yes	No	No	Yes
21	Yes	No	Yes	No	No	No	No
30	Yes	Yes	Yes	Yes	Yes	No	Yes
36	Yes	Yes	Yes	No	Yes	Yes	No
41	Yes	No	No	No	No	No	No
54	Yes	Yes	Yes	No	Yes	Yes	No
60	Yes	Yes	Yes	Yes	Yes	No	Yes
77	Yes	No	No	No	No	No	No

Number	1 times table	2 times table	3 times table	5 times table	6 times table	9 times table	10 times table
90	Yes	Yes	Yes	Yes	Yes	Yes	Yes
102	Yes	Yes	Yes	No	Yes	No	No
151	Yes	No	No	No	No	No	No
366	Yes	Yes	Yes	No	Yes	No	No

Task 4 Tests for divisibility 2

The 4 times table: Numbers in the 4 times table are every second number in the 2 times table. Also, 4 will divide into the last 2 digits of the number. Or halve the number and halve it again. If the answer is a whole number, the original number must be in the 4 times table.

The 8 times table: Numbers in the 8 times table are every second number in the 4 times table. Halve the number, halve the number and halve it again. If you have a whole number left, the original number will be in the 8 times table.

The 12 times table: Numbers in the 12 times table are in both the 3 and the 4 times tables.

The 11 times table: Numbers in the 11 times table have their odd-placed digits adding up to their even placed digits or differing by a number in the 11 times table ($165 = 1 + 5 = 6$).

The 7 times table: To see if a number is in the 7 times table, double the ones place and take it away from the other digits. The answer will be 0 or a number in the 7 times table ($84 = 8 - 4 \times 2 = 8 - 8 = 0$).

Number	1 times table	2 times table	3 times table	4 times table	5 times table	6 times table	7 times table	8 times table	9 times table	10 times table	11 times table	12 times table
12	Yes	Yes	Yes	Yes	No	Yes	No	No	No	No	No	Yes
29	Yes	No	No	No	No	No	No	No	No	No	No	No
36	Yes	Yes	Yes	Yes	No	Yes	No	No	Yes	No	No	Yes
54	Yes	Yes	Yes	No	No	Yes	No	No	Yes	No	No	No
60	Yes	Yes	Yes	Yes	Yes	Yes	No	No	No	Yes	No	Yes
73	Yes	No	No	No	No	No	No	No	No	No	No	No
81	Yes	No	Yes	No	No	No	No	No	Yes	No	No	No
100	Yes	Yes	No	No	Yes	No	No	No	No	Yes	Yes	No
210	Yes	Yes	Yes	No	Yes	Yes	Yes	No	No	Yes	No	No

A common multiple of the numbers 1 to 12 can be calculated by: $1 \times 2 \times 3 \times 4 \times 5 \times 6 \times 7 \times 8 \times 9 \times 10 \times 11 \times 12 = 479\,001\,600$.

However, because many of these numbers have common factors, the first number that is in all 12 tables is just 27 720. Every multiple of the 27 720 times table will be a common multiple of all tables from 1 to 12. It turns out that 479 001 600 is the 17 280th number that is in all 12 tables!

Task 5 Sports scores

Australian rules football: Geelong 71 beat West Coast Eagles 70.

Cricket: India 187 beat South Africa 160.

Rugby Union: England 47 drew with France 47.

Basketball: Detroit Pistons 101 beat Chicago Bulls 96.

Gaelic football: Cork 20 beat Clontarf 19.

Task 6 Multiplying big numbers

1.

Question	Rounded question	Rounded answer	Exact answer
54 × 67	50 × 70	3500	3618
21 × 67	20 × 70	1400	1407
55 × 49	50 × 50	2500	2695
378 × 4	400 × 4	1600	1512
6 × 923	6 × 900	5400	5538
109 × 43	110 × 40	4400	4687
112 × 123	110 × 120	13 200	13 776
226 × 589	200 × 600	120 000	133 114
2102 × 6	2000 × 6	12 000	12 612
7989 × 21	8000 × 20	160 000	167 769
1076 × 268	1100 × 300	330 000	288 368

The last rounded answer is a bit further away because both parts of the question have been rounded up.

2. Rounded answer for show entry: Adults: $20 × 4 = $80, Children: $5 × 3 = $15, Total = $95.
Cost of show entry: Adults: $20.40 × 4 = $81.60, Children: $4.90 × 3 = $14.70,
Total = $96.30.

3. Rounded income: 12 000 000 × $80 000 = $960 000 000 000.
Actual income: $960 569 444 214. (This estimate is better than $\frac{999}{1000}$ of the exact answer.)

Task 7 Short cuts

Multiplying by 2: 13 = 26, 24 = 48, 53 = 106, 106 = 212, 234 = 468, 1124 = 2248, 2507 = 5014.

Multiplying by 4: 23 = 92, 54 = 216, 72 = 288, 112 = 448, 345 = 1380, 1345 = 5380, 2235 = 8940.

Multiplying by 5: 14 = 70, 26 = 130, 45 = 225, 62 = 310, 123 = 615, 344 = 1720, 1456 = 7280.

Multiplying by 9: 13 = 117, 24 = 216, 37 = 333, 58 = 522, 120 = 1080, 230 = 2070, 561 = 5049.

Multiplying by 11: 14 = 154, 21 = 231, 36 = 396, 47 = 517, 68 = 748, 103 = 1133, 241 = 2651.

OXFORD UNIVERSITY PRESS

Left to right multiplication:

$32 \times 4 = (30 \times 4) + (2 \times 4) = 128.$

$53 \times 7 = (50 \times 7) + (3 \times 7) = 371.$

$59 \times 12 = (50 \times 12) + (9 \times 12) = 708.$

$42 \times 6 = (40 \times 6) + (2 \times 6) = 252.$

$66 \times 3 = (60 \times 3) + (6 \times 3) = 198.$

$82 \times 11 = (80 \times 11) + (2 \times 11) = 902.$

Task 8 Common multiples

1. 5: 20, 25, 50. 11: 44, 55, 110. 3: 12, 15, 30. 12: 48, 60, 120. 8: 32, 40, 80. 20: 80, 100, 200.

2. 2: 6, 18, 60. 4: 12, 36, 120. 6: 18, 54, 180. 7: 21, 63, 210. 9: 27, 81, 270. 10: 30, 90, 300.

3. 2 and 3: 6, 12, 18. 4 and 6: 12, 24, 36. 5 and 7: 35, 70, 105. 2 and 3 and 4: 12, 24, 36. 2 and 4 and 5: 20, 40, 60.

4. The lowest common multiple of 3, 4 and 5 is 60, so the 3 lighthouses will flash together every 60 minutes. The next time they will all flash together will be 1:30pm.

Task 9 The Sieve of Eratosthenes

The prime numbers to 150 are: 2, 3, 5, 7, 11, 13, 17, 19, 23, 29, 31, 37, 41, 43, 47, 53, 59, 61, 67, 71, 73, 79, 83, 89, 97, 101, 103, 107, 109, 113, 127, 131, 137, 139 and 149.

Thirty-five numbers are prime and 115 are not prime. (With the exception of 1, these are called composite numbers.)

Task 10 Problems to solve

1. The lowest common multiple of 4, 5 and 6 is 60, so they will all see each other at the library every 60 days. From 1 June they will be together next on '61 June' which is 31 July.

2. The lowest common multiple of 2, 3, 5 and 7 is 210. The fewest number of cards Sam could own would be 211.

3. The closer the garden bed is to a square the larger the area. A square is a subset of rectangles so the biggest area would be a 6 m × 6 m shape with an area of 36 sq m.

4. The digits 41 × 32 will give the biggest possible product of 1312.

5. The 3 tables scores are: 3 goals and 9 behinds (3 × 9 = 27), 4 goals and 8 behinds (4 × 8 = 32) and 7 goals and 7 behinds (7 × 7 = 49).

UNIT 6 – DIVIDE AND CONQUER
Division

Division is one of the four operations, alongside addition, subtraction and multiplication. All these operations are interrelated, meaning that they have much in common. Division is the inverse of multiplication ($6 \times 4 = 24$ so $24 \div 4 = 6$). Division is also repeated subtraction: $45 \div 9$ really means how many times can you take 9 away from 45 ($45 - 9 = 36$, $36 - 9 = 27$, $27 - 9 = 18$, $18 - 9 = 9$, $9 - 9 = 0$).

Research into how students learn the four operations has clearly demonstrated that students come to understand division much later than addition, subtraction and multiplication. This is partly due to the reliance of the division process on other mathematical concepts (multiplication, repeated subtraction), as well as the fact that we do not use it nearly as much in everyday life as the other three operations. Thus, compared to the other operations, you are recommended to use a slower approach to the teaching of division.

Keep division questions in relevant contexts. Use concrete materials to help less-able students. Relate division problems to multiplication, to show that the solution, multiplied by the divisor, must equal what is being divided in the first place. Brainstorming sessions should refer to sharing, breaking into equal groups, fractions, decimals, percentages, pizzas, money, games of sport, the clock face, time units and so on.

TASK 1
What does it mean?
Understanding the concept of division

There are lots of words and symbols in maths that mean 'to divide'. Can you think of some?

Discuss examples of when and why we use division. Alternatively, encourage students to draw a picture of something they might divide up at home or at school.

TASK 2
Equation creation
Connecting the four operations
You will need:

• cards numbered 1 to 10

Remind students that the four operations (addition, subtraction, multiplication and division) are very closely connected. Give the students a set of cards numbered 1 to 10.

OXFORD UNIVERSITY PRESS

Select two different cards and use them to write eight different equations – two addition equations, two subtraction equations, two multiplication equations and two division equations. For example:

> 2 and 8
> 2 + 8 = 10 and 8 + 2 = 10
> 10 – 2 = 8 and 10 – 8 = 2
> 2 × 8 = 16 and 8 × 2 = 16
> 16 ÷ 2 = 8 and 16 ÷ 8 = 2

TASK 3
Take it away
Using repeated subtraction
You will need:

*   counters
*   BLM 25

Remind students that multiplication is just a quick way of adding up. For example, 3 × 4 means 3 groups of 4, and is the same as adding up 4 three times (4 + 4 + 4 =12). In the same way, division is just a quick way of doing subtraction. For example, 12 ÷ 3 really means how many times can you subtract 3 from 12 before you reach 0 (12 – 3, – 3, – 3, – 3 = 0), with 4 as the answer.

Use counters to help turn division questions into subtraction questions. The first one has been done for you.

Question	Division as subtraction	Answer
30 ÷ 5	30 – 5 = 25 (1) 25 – 5 = 20 (2) 20 – 5 = 15 (3) 15 – 5 = 10 (4) 10 – 5 = 5 (5) 5 – 5 = 0 (6)	6
30 ÷ 6		
30 ÷ 10		
30 ÷ 3		
28 ÷ 4		
28 ÷ 7		
25 ÷ 5		
24 ÷ 12		
24 ÷ 8		
24 ÷ 6		
24 ÷ 4		
24 ÷ 3		
24 ÷ 2		

Question	Division as subtraction	Answer
20 ÷ 5		
20 ÷ 4		
18 ÷ 6		
18 ÷ 3		
16 ÷ 8		
16 ÷ 4		
16 ÷ 2		

TASK 4

Split it up

Sharing and grouping

You will need:

- counters
- BLM 26

Explain to students that when we divide numbers, we do not always get equal groups. For example, 24 can be split into 1, 2, 3 or 4 equal groups. But if we try to split it into 5 equal groups, there will be 4 left over. Show the students how to write the equation '24 divided by 5' with the answer as '4 "r 4"' to show the remainder. Remind the students that knowing their times tables will help them to work out whether a number will have a remainder or not when it is divided.

Use counters to split up 30, 24, 15 and 21 into groups. Record this as a division equation. Two examples have been done for you.

Number of counters: 30

Equal groups of ...	Yes or No	Equation	Equal groups of ...	Yes or No	Equation
5	Yes	30 ÷ 5 = 6	10		
3			4		
6			12	No	30 ÷ 12 = 2 r 6

TASK 5

What's left over

Recognising remainders

You will need:

- BLM 27

When we divide numbers, we do not always get a whole number answer. For example, 24 can be split into one, two, three or four equal groups. But when trying to split it into five equal groups, we encounter a 'leftover' number, or a remainder. Remind students that knowing their times tables will help them to work out whether a number will have a remainder when it is divided.

See if you can follow these tips to quickly fill out these 'Yes or No remainder' tables.

1. Can these numbers be divided by 2 with no remainder? (Remember that numbers in the 2 times table will always be even.)

Number	Yes or No	Remainder	Number	Yes or No	Remainder
12			31		
65			98		
107			160		
241			584		

2. Can these numbers be divided by 3 with no remainder? (Remember that numbers in the 3 times table have their digits summing to 3, 6 or 9.)

Number	Yes or No	Remainder	Number	Yes or No	Remainder
27			36		
71			80		
102			141		
223			460		

3. Can these numbers be divided by 5 with no remainder? (Remember that numbers in the 5 times table will end in 5 or 0.)

Number	Yes or No	Remainder	Number	Yes or No	Remainder
30			85		
97			115		
169			202		
370			406		

4. Can these numbers be divided by 9 with no remainder? (Remember that numbers in the 9 times table will have their digits summing to 9.)

Number	Yes or No	Remainder	Number	Yes or No	Remainder
27			45		
56			73		
96			117		
233			369		

TASK 6
Factor find
Identifying factors

Remind students that multiplying two whole numbers gives a product. The numbers that we multiply are the factors of the product. So for the number 9, we can see that 3, 1 and 9 are the factors: $1 \times 9 = 9$, $3 \times 3 = 9$.

We can use our times table knowledge to help find factors. For example, 9 can be found in the 1-, 3- and 9-times tables, so we know that 9 has 3 factors: 1, 3 and 9 ($1 \times 9 = 9$, $3 \times 3 = 9$, $9 \times 1 = 9$). The number 10 can be found in the 1, 2, 5 and 10 times tables, so we know that 10 has 4 factors: 1, 2, 5 and 10 ($1 \times 10 = 10$, $2 \times 5 = 10$, $5 \times 2 = 10$, $10 \times 1 = 10$).

1. See if you can find all the factors of the following numbers:

6	10	12	17	20	24	30	36
42	47	54	60	64	81	84	100

2. The number 6 is called a 'perfect number' because its factors (1, 2, 3, 6) add up to twice itself ($1 + 2 + 3 + 6 = 12$). Perfect numbers are very rare. There is only one 2-digit perfect number. Can you find it?

3. When a pair or group of numbers have the same factor or factors, these are called common factors. For example, the factors of 2 are 1 and 2, and the factors of 6 are 1, 2, 3 and 6; so 2 and 6 have common factors of 1 and 2. Can you find the common factors of these numbers?

   4 and 12

   8 and 24

   20 and 36

   50 and 100

TASK 7

Prime, composite or perfect square?
Using division to classify numbers

You will need:

- BLM 28

The more we know about mathematics, the more we can group and classify numbers. For example, numbers can be classified as odd or even, or whole numbers or rational numbers. By using division and our knowledge of factors, we can also classify numbers as being prime or composite.

Remind students that prime numbers have only two factors, while composite numbers have more than two factors. Perfect squares, with the exception of 1, are composite numbers, and they have an odd number of factors. When you list the factors of perfect squares in numerical order, the middle factor will be that number's square root.

For example, 9 is a composite number and a perfect square. Its factors are 1, 3 and 9. So the middle factor is the square root of 9: $3 \times 3 = 9$. A perfect square, like 9, is special because with 9 squares of the same size you can make a square.

Can you classify these numbers as being either prime (P), composite (C) or perfect square (PS)?

Number	P/ C/ PS	Number	P/ C/ PS	Number	P/ C/ PS	Number	P/ C/ PS	Number	P/ C/ PS
1		2		3		4		5	
6		7		8		9		10	
11		12		13		14		15	
16		17		18		19		20	
21		22		23		24		25	
26		27		28		29		30	
31		32		33		34		35	
36		37		38		39		40	
41		42		43		44		45	
46		47		48		49		50	
51		52		53		54		55	
56		57		58		59		60	
61		62		63		64		65	
66		67		68		69		70	
71		72		73		74		75	
76		77		78		79		80	
81		82		83		84		85	
86		87		88		89		90	
91		92		93		94		95	
96		97		98		99		100	

TASK 8

Finding common fractions

Applying division to fractions

You will need:

- BLM 29

When we say find one-third of an amount of money we are dealing with division rather than with fractions. What this question really means is 'divide the amount of money by 3'. When we find two-thirds of a number we need to divide that number by 3 and then double that answer. When we find three-quarters of a number, we need to divide that number by 4 and then triple that answer. And so on …

Can you use division to find the answers to the following questions found in these tables?

Can you use division to find one-half of	Answer	Can you use division to find one-third of	Answer	Can you use division to find one-quarter of	Answer
16		12		8	
28		21		20	
46		36		36	
78		48		48	
122		69		56	

Can you use division to find two-thirds of	Answer	Can you use division to find three-quarters of	Answer	Can you use division to find two-fifths of	Answer
12		12		5	
18		24		20	
27		36		35	
45		48		50	
120		60		80	

Can you use division to find three-eighths of	Answer	Can you use division to find five-sixths of	Answer	Can you use division to find eleven-twelfths of	Answer
16		12		12	
40		30		36	
56		42		72	
72		54		108	
160		78		240	

Can you use division to find one-and-a-half lots of	Answer	Can you use division to find one-and-two-third lots of	Answer	Can you use division to find one-and-three-quarter lots of	Answer
14		15		16	
22		24		32	
38		39		80	
52		51		112	
102		75		264	

TASK 9

Problems to solve

Applying division skills to unfamiliar situations

1. What is the first number that has factors of 1, 2, 3, 4, 5 and 6?
2. A number when divided by 6 gives the same result as when 30 is taken away from it. What is that number?
3. My Pokémon cards can be divided into 3, 4, 5 and 7 equal groups, with 2 cards left over on each occasion. I have fewer than 500 Pokémon cards. How many Pokémon cards do I have?
4. What is the first number that when divided by 5 and then by 5 again gives a whole number answer whose 2 digits sum to 10?
5. A, B, C, D and E are letters that stand for 5 different digits.

   When the 3-digit number CDE is divided by the 2-digit number AB, the answer is the 2-digit number 3B.

   What is the smallest digit that the letter B could stand for?

ANSWERS

Task 1 What does it mean?

Division words and symbols: divide, dividend, divisor, remainder, cut, split, separate, equal groups, quotient, fractions, decimals, percentage, share, etc.

Examples of use of division: when cutting up sandwiches, fruit, cakes or pizzas, when sharing amounts of money or cards or toys, when being fair with our friends, when splitting time units like hours or days up into equal parts, when dividing games up into sections, etc.

Task 2 Equation creation

Answers will vary.

Task 3 Take it away

Question	Division as subtraction	Answer
30 ÷ 5	(1) 25 (2) 20 (3) 15 (4) 10 (5) 5 (6) 0	6
30 ÷ 6	(1) 24 (2) 18 (3) 12 (4) 6 (5) 0	5
30 ÷ 10	(1) 20 (2) 10 (3) 0	3
30 ÷ 3	(1) 27 (2) 24 (3) 21 (4) 18 (5) 15 (6) 12 (7) 9 (8) 6 (9) 3 (10) 0	10
28 ÷ 4	(1) 24 (2) 20 (3) 16 (4) 12 (5) 8 (6) 4 (7) 0	7
28 ÷ 7	(1) 21 (2) 14 (3) 7 (4) 0	4

Question	Division as subtraction	Answer
25 ÷ 5	(1) 20 (2) 15 (3) 10 (4) 5 (5) 0	5
24 ÷ 12	(1) 12 (2) 0	2
24 ÷ 8	(1) 16 (2) 8 (3) 0	3
24 ÷ 6	(1) 18 (2) 12 (3) 6 (4) 0	4
24 ÷ 4	(1) 20 (2) 16 (3) 12 (4) 8 (5) 4 (6) 0	6
24 ÷ 3	(1) 21 (2) 18 (3) 15 (4) 12 (5) 9 (6) 6 (7) 3 (8) 0	8
24 ÷ 2	(1) 22 (2) 20 (3) 18 (4) 16 (5) 14 (6) 12 (7) 10 (8) 8 (9) 6 (10) 4 (11) 2 (12) 0	12
20 ÷ 5	(1) 15 (2) 10 (3) 5 (4) 0	4
20 ÷ 4	(1) 16 (2) 12 (3) 8 (4) 4 (5) 0	5
18 ÷ 6	(1) 12 (2) 6 (3) 0	3
18 ÷ 3	(1) 15 (2) 12 (3) 9 (4) 6 (5) 3 (6) 0	6
16 ÷ 8	(1) 8 (2) 0	2
16 ÷ 4	(1) 12 (2) 8 (3) 4 (4) 0	4
16 ÷ 2	(1) 14 (2) 12 (3) 10 (4) 8 (5) 6 (6) 4 (7) 2 (8) 0	8

Task 4 Split it up

Number of counters: 30

Equal groups of	Yes or No	Equation	Equal groups of	Yes or No	Equation
5	Yes	30 ÷ 5 = 6	10	Yes	30 ÷ 10 = 3
3	Yes	30 ÷ 3 = 10	4	No	30 ÷ 4 = 7 r 2
6	Yes	30 ÷ 6 = 5	12	No	30 ÷ 12 = 2 r 6

Number of counters: 24

Equal groups of	Yes or No	Equation	Equal groups of	Yes or No	Equation
5	No	24 ÷ 5 = 4 r 4	10	No	24 ÷ 10 = 2 r 4
3	Yes	24 ÷ 3 = 8	4	Yes	24 ÷ 4 = 6
6	Yes	24 ÷ 6 = 4	12	Yes	24 ÷ 12 = 2

Number of counters: 15

Equal groups of	Yes or No	Equation	Equal groups of	Yes or No	Equation
5	Yes	15 ÷ 5 = 3	10	No	15 ÷ 10 = 1 r 5
3	Yes	15 ÷ 3 = 5	4	No	15 ÷ 4 = 3 r 3
6	No	15 ÷ 6 = 2 r 3	12	No	15 ÷ 12 = 1 r 3

Number of counters: 21

Equal groups of	Yes or No	Equation	Equal groups of	Yes or No	Equation
5	No	21 ÷ 5 = 4 r 1	10	No	21 ÷ 10 = 2 r 1
3	Yes	21 ÷ 3 = 7	4	No	21 ÷ 4 = 5 r 1
6	No	21 ÷ 6 = 3 r 3	12	No	21 ÷ 12 = 1 r 9

Task 5 Leftovers

1.

Number	Yes or No	Remainder	Number	Yes or No	Remainder
12	Yes	0	31	No	1
65	No	1	98	Yes	0
107	No	1	160	Yes	0
241	No	1	584	Yes	0

2.

Number	Yes or No	Remainder	Number	Yes or No	Remainder
27	Yes	0	36	Yes	0
71	No	2	80	No	2
102	Yes	0	141	Yes	0
223	No	1	460	No	1

3.

Number	Yes or No	Remainder	Number	Yes or No	Remainder
30	Yes	0	85	Yes	0
97	No	2	115	Yes	0
169	No	4	202	No	2
370	Yes	0	406	No	1

4.

Number	Yes or No	Remainder	Number	Yes or No	Remainder
27	Yes	0	45	Yes	0
56	No	2	73	No	1
96	No	6	117	Yes	0
233	No	8	369	Yes	0

Task 6 Factor find

1. 6: 1, 2, 6.

10: 1, 2, 5, 10.

12: 1, 2, 3, 4, 6, 12.

17: 1, 17.

20: 1, 2, 4, 5, 10, 20.

24: 1, 2, 3, 4, 6, 8, 12, 24.

30: 1, 2, 3, 5, 6, 10, 15, 30.

36: 1, 2, 3, 4, 6, 9, 12, 18, 36.

42: 1, 2, 3, 6, 7, 14, 21, 42.

47: 1, 47.

54: 1, 2, 3, 6, 9, 18, 27, 54.

60: 1, 2, 3, 4, 5, 6, 10, 12, 15, 20, 30, 60.

64: 1, 2, 4, 8, 16, 32, 64.

81: 1, 3, 9, 27, 81.

84: 1, 2, 3, 4, 6, 7, 12, 14, 21, 28, 42, 84.

100: 1, 2, 4, 5, 10, 20, 25, 50, 100.

2. The second perfect number and the only 2-digit number that is a perfect number is 28 (1 + 2 + 4 + 7 + 14 + 28 = 56).

3. 4 and 12: 1, 2 and 4.

8 and 24: 1, 2, 4 and 8.

20 and 36: 1, 2 and 4.

50 and 100: 1, 2, 5, 10, 25 and 50.

Task 7 Prime, composite or perfect square?

Number	P/ C/ PS	Number	P/ C/ PS	Number	P/ C/ PS	Number	P/ C/ PS	Number	P/ C/ PS
1	PS	2	P	3	P	4	PS	5	P
6	C	7	P	8	C	9	PS	10	C
11	P	12	C	13	P	14	C	15	C
16	PS	17	P	18	C	19	P	20	C
21	C	22	C	23	P	24	C	25	PS
26	C	27	C	28	C	29	P	30	C
31	P	32	C	33	C	34	C	35	C
36	PS	37	P	38	C	39	C	40	C
41	P	42	C	43	P	44	C	45	C
46	C	47	P	48	C	49	PS	50	C
51	C	52	C	53	P	54	C	55	C
56	C	57	C	58	C	59	P	60	C
61	P	62	C	63	C	64	PS	65	C
66	C	67	P	68	C	69	C	70	C
71	P	72	C	73	P	74	C	75	C
76	C	77	C	78	C	79	P	80	C
81	PS	82	C	83	P	84	C	85	C

OXFORD UNIVERSITY PRESS

Number	P/ C/ PS	Number	P/ C/ PS	Number	P/ C/ PS	Number	P/ C/ PS	Number	P/ C/ PS
86	C	87	C	88	C	89	P	90	C
91	C	92	C	93	C	94	C	95	C
96	C	97	P	98	C	99	C	100	PS

Task 8 Finding common fractions

One-half of	Answer	One-third of	Answer	One-quarter of	Answer
16	8	12	4	8	2
28	14	21	7	20	5
46	23	36	12	36	9
78	39	48	16	48	12
122	61	69	23	56	14

Two-thirds of	Answer	Three-quarters of	Answer	Two-fifths of	Answer
12	8	12	9	5	2
18	12	24	18	20	8
27	18	36	27	35	14
45	30	48	36	50	20
120	80	60	45	80	32

Three-eighths of	Answer	Five-sixths of	Answer	Eleven-twelfths of	Answer
16	6	12	10	12	11
40	15	30	25	36	33
56	21	42	35	72	66
72	27	54	45	108	99
160	60	78	65	240	220

One-and-a-half lots of	Answer	One-and-two-third lots of	Answer	One-and-three-quarter lots of	Answer
14	21	15	25	16	28
22	33	24	40	32	56
38	57	39	65	80	140
52	78	51	85	112	196
102	153	75	125	264	462

Task 9 Problems to solve

1. 60 is the smallest number that has 1 to 6 as factors.
2. 36 divided by 6 equals 36 − 30.
3. 3, 4, 5 and 7 can be divided equally into 420. Add 2 and the answer is 422.
4. 475 divided by 5 equals 95. 95 divided by 5 equals 19. 1 + 9 = 10.
5. Let AB equal 12 and CDE equal 384. This gives the result of 32. The smallest value B can equal is 2.

UNIT 7 – A FRACTION FEAST
Fractions

A study of fractions should follow on logically from a study of division, as fractions are merely applied division problems. For example, when we are dealing with three-quarters of an object, it literally means to divide that object into four equal pieces and then to take three of them.

A fraction is a ratio (a comparison of one quantity compared to another) and comes under the mathematical realm of rational numbers. When forming fractions equal to one half for example, the answers must be in the form of one part to two parts. Eight-sixteenths must be equal to one half because the ratio of the numerator to the denominator in both instances is 1 part to 2 parts (1:2 = 8:16).

Forming equal fractions is a key component of rational number work and it is this reference to ratios that is so important in fostering students' deep understanding of fractions.

Use many diverse sets of concrete materials when teaching fractions, such as fraction circles, strips and fraction walls. When brainstorming, refer to fractions found in pizzas and other foods, sporting games, sharing, the sorting of crowds into adults/children, tallies and statistical analysis of graphs. And remember, whenever we deal with probability (e.g. rolling dice, cutting cards, tossing a coin or spinning a spinner) we are dealing with probability, and the chances of events occurring in games of probability are often expressed as fractions.

I Whole

| $\frac{1}{2}$ | $\frac{1}{2}$ |

| $\frac{1}{3}$ | $\frac{1}{3}$ | $\frac{1}{3}$ |

| $\frac{1}{4}$ | $\frac{1}{4}$ | $\frac{1}{4}$ | $\frac{1}{4}$ |

| $\frac{1}{5}$ | $\frac{1}{5}$ | $\frac{1}{5}$ | $\frac{1}{5}$ | $\frac{1}{5}$ |

| $\frac{1}{6}$ | $\frac{1}{6}$ | $\frac{1}{6}$ | $\frac{1}{6}$ | $\frac{1}{6}$ | $\frac{1}{6}$ |

| $\frac{1}{8}$ | $\frac{1}{8}$ | $\frac{1}{8}$ | $\frac{1}{8}$ | $\frac{1}{8}$ | $\frac{1}{8}$ | $\frac{1}{8}$ | $\frac{1}{8}$ |

| $\frac{1}{10}$ | $\frac{1}{10}$ | $\frac{1}{10}$ | $\frac{1}{10}$ | $\frac{1}{10}$ | $\frac{1}{10}$ | $\frac{1}{10}$ | $\frac{1}{10}$ | $\frac{1}{10}$ | $\frac{1}{10}$ |

| $\frac{1}{12}$ | $\frac{1}{12}$ | $\frac{1}{12}$ | $\frac{1}{12}$ | $\frac{1}{12}$ | $\frac{1}{12}$ | $\frac{1}{12}$ | $\frac{1}{12}$ | $\frac{1}{12}$ | $\frac{1}{12}$ | $\frac{1}{12}$ | $\frac{1}{12}$ |

TASK 1

Fractions everywhere

Understanding the concept of fractions

You will need:

- BLM 30
- circle divided into four parts

1. What do you think a fraction is?

2. Where can you see fractions at school or at home?

Remind students that when we write fractions, they have three parts: the bottom part is called the denominator and tells us into how many equal parts something has been divided into; the top is called the numerator and tells us how many of these equal parts we are using. The linc that separates the numerator and the denominator is called a 'vinculum' or 'division bar'.

Show students a circle that has been divided up into four equal parts and show them the denominator. Indicate one part to show the numerator of $\frac{3}{4}$, and point out the vinculum.

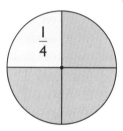

3. Shade $\frac{1}{4}$ of these shapes.

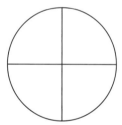  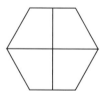

TASK 2

Kitchen fractions

Making and writing common fractions

You will need:

- glass
- access to water
- sandwich
- knife
- loaf of bread
- apple

1. Fill a glass half full of water. Draw and label the fraction you have made.
2. Make a sandwich and cut it into quarters. Draw and label the fractions you have made.
   If you ate three-quarters of the sandwich, what fraction of the sandwich would be left?
3. I use one-third of a cup of milk when I make my porridge in the microwave each morning.
   Draw one-third of a cup of milk. What fraction of the glass is empty when one-third of it is milk?
4. Count and draw the slices in a loaf of bread. What fraction of the loaf is 1 slice of bread?
5. Cut an apple in half and draw each piece. Now cut each of the two pieces in half and draw them. Now cut each of the four pieces in half and draw what you see.
   What fraction of the apple is one of these pieces?

TASK 3

A fraction symphony

Measuring fractions

You will need:

- nine glasses
- water
- spoon or pencil

Sound travels through water at different speeds, depending on the depth of the water. When you tap a glass with water in it the water molecules vibrate, which creates sound waves. In this task students will be able to prove this point by making an orchestra using glasses and water.

Use nine glasses of exactly the same size. Leave the first glass empty.
Now fill one glass to half full.
Now fill one glass to half that amount.
Now fill one glass to half that amount again. Now fill one glass so that it is full of water. Now fill one to halfway between half and full.
Now line up these six glasses in order, from empty to full, with the empty glass on the left and the full glass on the right.
Now fill a glass so that its water will be halfway between the 3rd and the 4th glass and put it in the correct order.
Now fill a glass so that its water will be halfway between the 5th and the 6th glass and put it in the correct order.
Finally fill a glass so that its water will be halfway between the 7th and the 8th glass and put it in the correct order.

1. What fraction full is each glass?
2. Can you think of a different and better way of saying some of these fractions, like $\frac{4}{8}$?
3. Now let's make some music! Get a spoon or a pencil and tap each glass from empty to full. What do you notice?
4. Play some simple tunes using your glass orchestra, such as 'Baa, Baa Black Sheep' or 'Twinkle, Twinkle Little Star'. Or make up your own watery fraction tune!

TASK 4

Fill to a mark

Estimating fractions in capacity

You will need:

- glass, cup or mug
- two different-sized saucepans
- bucket
- measuring jug
- calculator

Get out glass, cup or mug, 2 different-sized saucepans, a bucket, a measuring jug and a calculator.

In this task, students will estimate what fraction of water is in a container or vessel. However, this time they will have the ability to check the accuracy of their estimates.

1. Use a measuring jug to find how much water a glass will hold in millilitres when it is full.

   If you need to, use a calculator to find out half of the capacity of the glass (divide by 2).

   First, use an unmarked jug to fill the glass until it is about half full. Now use the measuring jug to see how close your estimate was.

   How close was your estimate?

2. Use a measuring jug to find how much water a cup or mug will hold in millilitres when it is full.

   If you need to, use a calculator to find out a quarter of the capacity of the cup (divide by 4).

   First use an unmarked jug to fill the cup until it is about a quarter full. Now use the measuring jug to see how close your estimate was.

   How close was your estimate?

3. Get a small saucepan. Use a measuring jug to find how much water it will hold in millilitres when it is full.

   If you need to, use a calculator to find out one-third of the capacity of the saucepan (divide by 3).

   Using just estimation, fill the saucepan so that you think it is one-third full. Now use the measuring jug to see how close your estimate was.

   How close was your estimate?

4. Get a larger saucepan. Use a measuring jug to find how much water it will hold in millilitres when it is full.

   If you need to, use a calculator to find out one-fifth of the capacity of the saucepan (divide by 5).

Using just estimation, fill the saucepan so that you think it is one-fifth full. Now use the measuring jug to see how close your estimate was.

How close was your estimate?

5. Get a bucket. Use a measuring jug to find how much water it will hold in millilitres when it is full.

If you need to, use a calculator to find out one-tenth of the capacity of the bucket.

Using only estimation, fill the bucket so that you think it is one-tenth full. Now use the measuring jug to see how close your estimate was.

How close was your estimate?

TASK 5

One to 10 then start again

Naming fractions

You will need:

- playing cards numbered 1 (ace) to 10
- BLM 31

Ask students to make fractions by turning over two different cards, using the smaller number as the numerator and the larger number as the denominator to create a simple or proper fraction (one or less).

Remind students that when a fraction is equal to 1 whole number, its numerator and denominator are the same, like $\frac{4}{4}$. In this example, something, like a game of netball, has been split up into 4 equal parts and all 4 have been used.

Shuffle the cards and choose two cards to make a fraction, with the smallest number at the top, and fill in the table. If you create a fraction that is already on this table, make another. There are 45 possible answers! Can you see why?

Numerator	Denominator	Fraction	Fraction for 1	Difference from 1 whole number
4	5	$\frac{4}{5}$	$\frac{5}{5}$	$\frac{1}{5}$

Numerator	Denominator	Fraction	Fraction for 1	Difference from 1 whole number

TASK 6

Footy fractions

Using fractions in the AFL

You will need:

- BLM 32

The Australian Football League (AFL) began in 1990 and this table shows the AFL premiers for the first 30 years. Remind students that a fraction in its simplest form uses the smallest numerator and denominator possible, by dividing by the biggest factor that these two numbers have in common. For example, $\frac{6}{30}$ – 6 is the biggest common factor so $\frac{6}{30}$ is $\frac{1}{5}$ in simplest form.

Use the table to work out what fraction of the AFL premierships each team has won, in its simplest form.

Year	Premier	Year	Premier	Year	Premier	Year	Premier	Year	Premier
1990	Collingwood	1991	Hawthorn	1992	West Coast Eagles	1993	Essendon	1994	West Coast Eagles
1995	Carlton	1996	North Melbourne	1997	Adelaide Crows	1998	Adelaide Crows	1999	North Melbourne
2000	Essendon	2001	Brisbane Lions	2002	Brisbane Lions	2003	Brisbane Lions	2004	Port Adelaide
2005	Sydney Swans	2006	West Coast Eagles	2007	Geelong	2008	Hawthorn	2009	Geelong

OXFORD UNIVERSITY PRESS

Year	Premier	Year	Premier	Year	Premier	Year	Premier	Year	Premier
2010	Collingwood	2011	Geelong	2012	Sydney Swans	2013	Hawthorn	2014	Hawthorn
2015	Hawthorn	2016	Western Bulldogs	2017	Richmond	2018	West Coast Eagles	2019	Richmond

Team	AFL premierships	Fraction won in simplest form	Team	AFL premierships	Fraction won in simplest form
Adelaide Crows			Hawthorn		
Brisbane Lions			Melbourne		
Carlton			North Melbourne		
Collingwood			Port Adelaide		
Essendon			Richmond		
Fremantle			St Kilda		
Geelong			Sydney Swans		
Gold Coast Suns			Western Bulldogs		
Greater Western Sydney			West Coast Eagles		

What would you get if you added up all 18 fractions?

TASK 7

Chances are

Connecting fractions and probability

You will need:

• playing cards

Explain to students that whenever we roll dice, cut cards, toss a coin or spin a spinner we are dealing with the area of maths called probability. Chances of events occurring in games of probability are often expressed as fractions.

See if you can work out the chance of the events in these tables occurring.

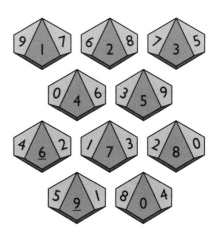

1. Rolling a 0 to 9-sided die

Event	Chance as a fraction	Event	Chance as a fraction
Rolling a 5		Rolling a number bigger than 6	
Rolling a number		Rolling an even number	
Rolling a prime number		Rolling an odd number	
Rolling a digit with an 'e' in its name		Rolling a number starting with a vowel	

2. Tossing a coin

Event	Chance as a fraction	Event	Chance as a fraction
Tossing a head		Tossing a tail	
Tossing 2 heads		Tossing 2 tails	
Tossing a head and then a tail		Tossing a tail and then a head	
Tossing 3 heads in a row		Tossing a tail and then a head and then a tail	

  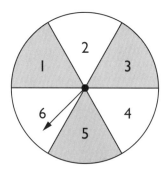

3. Spinning a 6-sectioned spinner

Event	Chance as a fraction	Event	Chance as a fraction
Spinning a 4		Spinning a 3 or a 4	
Spinning a number less than 5		Spinning a number	
Spinning a prime number		Spinning a number with 3 letters in its name	
Spinning a number with an 'a' in its name		Spinning a number in the 2 or the 3 times tables	

OXFORD UNIVERSITY PRESS

4. Cutting a 52-card deck of playing cards

Event	Chance as a fraction	Event	Chance as a fraction
Picking a red card		Picking a black card	
Picking a spade		Picking a heart	
Picking a 3		Picking a jack or queen	
Picking the 4 of spades		Picking a number	

TASK 8

Fraction folding

Relating fractions to perimeter and area

You will need:

- A4 paper
- scissors

Give students a piece of A4 paper and ask them to use scissors to trim it so that it is 24 cm long and 16 cm wide.

You will need to fold this piece of paper in half, in half and in half again. There are four completely different ways of doing this. Can you find them all? Draw the rectangle created by each method and label the length, width, perimeter and area.

What do you notice about the perimeters of the 4 rectangles?

What did you notice about the areas of the 4 rectangles?

TASK 9

Matching fractions

Connecting fractions, decimals and percentages

You will need:

- BLM 33
- calculator

Explain to students that fractions, decimals and percentages can all be shown as a ratio of a numerator to a denominator – they are all rational numbers. Remember that the line that separates the numerator and the denominator of a fraction is called a 'vinculum' or 'division bar'.

1. Any fraction can be turned into a decimal by thinking of the fraction as a division sum (which is exactly what it is). For example, to turn $\frac{4}{5}$ into a decimal you can use a calculator to work out $4 \div 5$. Round your answer, if necessary, to three decimal places.

Fraction	Decimal	Fraction	Decimal
$\frac{1}{2}$		$\frac{1}{4}$	
$\frac{3}{4}$		$\frac{1}{10}$	
$\frac{7}{10}$		$\frac{1}{5}$	
$\frac{3}{5}$		$\frac{1}{8}$	
$\frac{7}{8}$		$\frac{1}{3}$	
$\frac{2}{3}$		$\frac{1}{20}$	
$\frac{19}{20}$		$\frac{17}{100}$	
$\frac{47}{100}$		$\frac{1}{50}$	
$\frac{17}{50}$		$\frac{1}{7}$	

2. Any fraction can be turned into a percentage by thinking of the fraction as a division sum and multiplying it by 100. For example, to change $\frac{2}{9}$ into a percentage you can use a calculator to work out $2 \div 9 \times 100$. Round your answer, if necessary, to three decimal places.

Fraction	Percentage	Fraction	Percentage
$\frac{1}{2}$		$\frac{1}{4}$	
$\frac{3}{4}$		$\frac{1}{10}$	
$\frac{7}{10}$		$\frac{1}{5}$	
$\frac{3}{5}$		$\frac{1}{8}$	
$\frac{7}{8}$		$\frac{1}{3}$	
$\frac{2}{3}$		$\frac{1}{20}$	
$\frac{19}{20}$		$\frac{17}{100}$	
$\frac{47}{100}$		$\frac{1}{50}$	
$\frac{17}{50}$		$\frac{1}{7}$	

3. What connection do you see between the two tables?

OXFORD UNIVERSITY PRESS

TASK 10
Winning big
Multiplying fractions to calculate possible outcomes

You will need:

* coin
* dice
* playing cards

Explain to students that many games of chance involve fractions. Using fractions in games of chance can tell us the number of possible outcomes and, in turn, the probability of winning.

Use fractions to calculate the odds of winning games of chance. The first example for each game has been done for you.

Win the toss

When tossing a coin, what is the chance of tossing a head four times in a row? List all the possible outcomes. The first one has been done for you.

1. HHHH   2. _____   3. _____   4. _____

5. _____   6. _____   7. _____   8. _____

9. _____   10._____   11._____   12._____

13._____   14._____   15._____   16._____

Because the chance of tossing a head is $\frac{1}{2}$ on each occasion, simply multiply $\frac{1}{2} \times \frac{1}{2} \times \frac{1}{2} \times \frac{1}{2}$ to get $\frac{1}{16}$: this is the chance of tossing a head 4 times in a row.

Roll 666

What is the chance of rolling a 6 on a 6-sided dice, three times in a row? List ten possible outcomes. The first one has been done for you.

1. 1, 1, 1   2. _____   3. _____   4. _____   5. _____

6. _____   7. _____   8. _____   9. _____   10._____

What is the problem with listing outcomes?

Can you multiply fractions to find out the chance _____ $= \frac{1}{\rule{1cm}{0.4pt}}$

Two aces

What would the chances be of shuffling a deck of cards, picking an ace, putting the card back, shuffling the deck and again picking another ace?

List 10 possible ways of picking 2 cards from a deck.

1. ace, ace   2. _____   3. _____   4. _____   5. _____

6. _____   7. _____   8. _____   9. _____   10. _____

Now use multiplication of fractions to find the chance _____ = $\frac{1}{\underline{}}$

123

What would the chance be of getting 10 cards numbered 1 to 10, shuffling them, picking a 1, putting the card back, shuffling the cards, picking a 2, putting the card back, shuffling the cards and then picking a 3?

List 10 possible outcomes.

1. 1, 1, 1   2. _____   3. _____   4. _____   5. _____

6. _____   7. _____   8. _____   9. _____   10. _____

Now use multiplication of fractions to find the chance _____ = $\frac{1}{\underline{}}$

Head then jack

What would the chance be of tossing a head and then picking a jack from a deck of cards?

List 10 possible outcomes.

1. H, ace   2. _____   3. _____   4. _____   5. _____

6. _____   7. _____   8. _____   9. _____   10. _____

Now use multiplication of fractions to find the chance _____ = $\frac{1}{\underline{}}$

TASK 11

Problems to solve

Applying fraction skills to unfamiliar situations

1. Can you find a fraction that is exactly halfway between $\frac{1}{5}$ and $\frac{1}{4}$?

2. On a train, one-half of the passengers are men and one-third are women. There are 30 children on the train. How many people are on the train in total?

3. The top AFL footballers can earn $1 000 000 a year. A football game is made up of four quarters. If a player is injured, and only plays 6 and $\frac{1}{4}$ games of football for the year, how much money, per quarter of football, do they earn?

4. The wealthy businesswoman Leafy Wood has just made her will. She has decided to leave half of her money to her husband, a quarter to her son, an eighth to her niece, a sixteenth to her butler and the remaining $500 000 to the Lost Dogs' Home.
   How much money does Leafy Wood have?

5. Grandma saves $1 and $2 coins in a jar and at Christmas time each year, she splits the money up and gives it to her grandchildren. Grandma notices that the money in the jar can be split into half with $1 left over, can be split into thirds with $1 left over, can be split into

OXFORD UNIVERSITY PRESS

quarters with $1 left over, can be split into fifths with $1 left over and can be split into sixths with $1 left over. Grandma has more than $200 in the jar.

What is the smallest amount of money Grandma could have in the jar?

ANSWERS

Task 1 Fractions everywhere

1. Answers will vary but could include: a part of something, the result of doing division, something cut into equal parts.
2. Answers will vary but may include: when cutting up pizzas, sandwiches or fruit; in games and sport to break up the match or contest into sections; when dealing with time and parts of the hour or day; when cooking and following recipes; when dealing with games of chance.
3. One of the 4 equal parts should be shaded.

Task 2 Kitchen fractions

1. $\frac{1}{2}$

2. $\frac{1}{4}$

3. $\frac{2}{3}$

4. Answers may vary. From $\frac{1}{20}$ to $\frac{1}{24}$

5. $\frac{1}{8}$

Task 3 A fraction symphony

1. $\frac{0}{8}, \frac{1}{8}, \frac{2}{8}, \frac{3}{8}, \frac{4}{8}, \frac{5}{8}, \frac{6}{8}, \frac{7}{8}, \frac{8}{8}$

2. $0, \frac{1}{8}, \frac{1}{4}, \frac{3}{8}, \frac{1}{2}, \frac{5}{8}, \frac{3}{4}, \frac{7}{8}, 1$

3. More water makes for slower vibration of the molecules and so the 'notes' will get lower.

Task 4 Fill to a mark

Answers will vary.

Task 5 One to 10 then start again

There are 45 possible answers because there are 9 possible answers with 1 as the numerator, 8 possible answers with 2 as the numerator, 7 with 3 as the numerator and so on, to only 1 with 9 as the numerator ($\frac{9}{10}$). The sum of 1 to 9 is 45.

Answers will vary but the numerators of the 2 fractions that sum to 1 whole number must together equal the denominator, like $\frac{2}{7}$ and $\frac{5}{7}$ (2 + 5 = 7).

Task 6 Footy fractions

Team	AFL premierships	Fraction won in simplest form	Team	AFL premierships	Fraction won in simplest form
Adelaide Crows	2	$\frac{1}{15}$	Hawthorn	5	$\frac{1}{6}$
Brisbane Lions	3	$\frac{1}{10}$	Melbourne	0	$\frac{0}{30}$
Carlton	1	$\frac{1}{30}$	North Melbourne	2	$\frac{1}{15}$
Collingwood	2	$\frac{1}{15}$	Port Adelaide	1	$\frac{1}{30}$
Essendon	2	$\frac{1}{15}$	Richmond	2	$\frac{1}{15}$
Fremantle	0	$\frac{0}{30}$	St Kilda	0	$\frac{0}{30}$
Geelong	3	$\frac{1}{10}$	Sydney Swans	2	$\frac{1}{15}$
Gold Coast Suns	0	$\frac{0}{30}$	Western Bulldogs	1	$\frac{1}{30}$
Greater Western Sydney	0	$\frac{0}{30}$	West Coast Eagles	4	$\frac{2}{15}$

The fractions will all add up to 1 whole number.

Task 7 Chances are

1. $\frac{1}{10}, \frac{3}{10}, \frac{9}{10}, \frac{4}{10} \left(\frac{2}{5}\right), \frac{4}{10} \left(\frac{2}{5}\right), \frac{5}{10} \left(\frac{1}{2}\right), \frac{7}{10}, \frac{2}{10} \left(\frac{1}{5}\right).$

2. $\frac{1}{2}, \frac{1}{2}, \frac{1}{4}, \frac{1}{4}, \frac{1}{4}, \frac{1}{4}, \frac{1}{8}, \frac{1}{8}.$

3. $\frac{1}{6}, \frac{2}{6} \left(\frac{1}{3}\right), \frac{4}{6} \left(\frac{2}{3}\right), \frac{6}{6} (1), \frac{3}{6} \left(\frac{1}{2}\right), \frac{3}{6} \left(\frac{1}{2}\right), \frac{0}{6} (0), \frac{4}{6} \left(\frac{2}{3}\right).$

4. $\frac{26}{52} \left(\frac{1}{2}\right), \frac{26}{52} \left(\frac{1}{2}\right), \frac{13}{52} \left(\frac{1}{4}\right), \frac{13}{52} \left(\frac{1}{4}\right), \frac{4}{52} \left(\frac{1}{13}\right), \frac{8}{52} \left(\frac{2}{13}\right), \frac{1}{52}, \frac{36}{52} \left(\frac{9}{13}\right).$

Task 8 Fraction folding

	Length	Width	Perimeter	Area of rectangle
Method 1	24 cm	2 cm	52 cm	48 sq cm
Method 2	16 cm	3 cm	38 cm	48 sq cm
Method 3	12 cm	4 cm	32 cm	48 sq cm
Method 4	8 cm	6 cm	28 cm	48 sq cm

OXFORD UNIVERSITY PRESS

The perimeters of the 4 rectangles are all different. The closer the rectangle is to a square the smaller the perimeter is.

The areas of the 4 rectangles are all the same.

Task 9 Matching fractions

1.

Fraction	Decimal	Fraction	Decimal
$\dfrac{1}{2}$	0.5	$\dfrac{1}{4}$	0.25
$\dfrac{3}{4}$	0.75	$\dfrac{1}{10}$	0.1
$\dfrac{7}{10}$	0.7	$\dfrac{1}{5}$	0.2
$\dfrac{3}{5}$	0.6	$\dfrac{1}{8}$	0.125
$\dfrac{7}{8}$	0.875	$\dfrac{1}{3}$	0.333
$\dfrac{2}{3}$	0.667	$\dfrac{1}{20}$	0.05
$\dfrac{19}{20}$	0.95	$\dfrac{17}{100}$	0.17
$\dfrac{47}{100}$	0.47	$\dfrac{1}{50}$	0.02
$\dfrac{17}{50}$	0.34	$\dfrac{1}{7}$	0.143

2.

Fraction	Percentage	Fraction	Percentage
$\dfrac{1}{2}$	50%	$\dfrac{1}{4}$	25%
$\dfrac{3}{4}$	75%	$\dfrac{1}{10}$	10%
$\dfrac{7}{10}$	70%	$\dfrac{1}{5}$	20%
$\dfrac{3}{5}$	60%	$\dfrac{1}{8}$	12.5%
$\dfrac{7}{8}$	87.5%	$\dfrac{1}{3}$	33.333%
$\dfrac{2}{3}$	66.667%	$\dfrac{1}{20}$	5%
$\dfrac{19}{20}$	95%	$\dfrac{17}{100}$	17%
$\dfrac{47}{100}$	47%	$\dfrac{1}{50}$	2%
$\dfrac{17}{50}$	34%	$\dfrac{1}{7}$	14.286%

3. The connection is that the percentages are the decimals multiplied by 100.

Task 10 Winning big

Win the toss: 1. HHHH 2. HHHT 3. HHTH 4. HTHH 5. HHTT 6. HTHT 7. HTTH 8. HTTT 9. TTTT 10. TTTH 11. TTHT 12. THTT 13. TTHH 14. THTH 15. THHT 16. THHH.

Roll 666: The problem is that there are so many possible outcomes it takes too long to list them all. $\frac{1}{6} \times \frac{1}{6} \times \frac{1}{6} = \frac{1}{216}$, the chance of rolling three 6s in a row.

Two aces: $\frac{1}{13} \times \frac{1}{13} = \frac{1}{169}$.

123: $\frac{1}{10} \times \frac{1}{10} \times \frac{1}{10} = \frac{1}{1000}$.

Head then jack: $\frac{1}{2} \times \frac{1}{13} = \frac{1}{26}$.

Task 11 Problems to solve

1. Find a common denominator (20). Change the fractions to $\frac{4}{20}$ and $\frac{5}{20}$. Double the denominators to 40, making the fractions $\frac{8}{40}$ and $\frac{10}{40}$. Halfway is $\frac{9}{40}$.

2. If $\frac{1}{2}$ are men on the train and $\frac{1}{3}$ are women, that leaves $\frac{1}{6}$ to be children. If 30 is $\frac{1}{6}$ of a number, that number must be 180.

3. There are 25 quarters in 6 and $\frac{1}{4}$ games of football. So $1 000 000 divided by 25 equals $40 000 per quarter.

4. $\frac{1}{2} + \frac{1}{4} + \frac{1}{8} + \frac{1}{16}$ together makes $\frac{15}{16}$. So the $500 000 going to the Lost Dogs' Home must represent $\frac{1}{16}$ of the inheritance. $500 000 × 16 = $8 000 000.

5. The amount of money in the jar must be a common multiple of 1, 2, 3, 4, 5 and 6 +1. The lowest common multiple of 1 to 6 that is greater than 200 is 240. 240 + 1 = $241.

UNIT 8 – DECIMAL DELIGHTS
Decimals

Any quantity that can be expressed as a fraction comes under the realm of rational numbers, therefore decimals (providing that they are not recurring like Pi and most square roots) are also rational numbers. In the past, decimals used to be called 'decimal fractions'. This was to show that decimals are really no more than fractions, or parts of something, using our Base 10 number system.

Students need a good understanding of the concept of fractions such as 'a tenth' or 'a hundredth' to understand how decimals work, and decimals should be introduced to students via the place value chart. We can use the place value chart to show students how our numbers get 10 times bigger (1, 10, 100, 1000, etc) as we move to the left on the place value chart, and how when we move to the right on the chart, the numbers get 10 times smaller (1000, 100, 10, 1). Explain that ten times smaller than 1 is $\frac{1}{10}$ or 0.1 and ten times smaller again is $\frac{1}{100}$ or 0.01, etc. You can use multi base arithmetic blocks (MABs) to show how decimals work: simply pretend that each block is ten times smaller in value than usual and the unit becomes equal to 0.1. Decrease by ten times again and the 10 block becomes 0.1 and the unit 0.01.

When brainstorming decimals, refer to: money, sporting events with decimal measurements like the long jump or high jump, times, food labels with nutrition information, temperatures, petrol prices and pie charts. Time is one area where decimals are used widely. When we time an activity, particularly in a competition or sporting context, we often need very accurate measures. Show the students how the stopwatch on a smartphone measures the passage of time in minutes, seconds and hundredths of a second, shown as a decimal. For example, 1:59.68 shows 1 minute and 59 point 68 seconds – the 'point 68' means 68 hundredths of the next second, so this time could also be seen as 32 hundredths of a second short of 2 minutes.

Sometimes we do not need an exact answer or numbers. Remind students that we can use rounding to make decimals easier to understand, or when an exact answer is not required, just as we often do for whole numbers or numbers with remainders. And just as is the case when rounding whole numbers, make sure students understand that 5 is the key digit to consider when rounding decimals. So decimals with .5 or more (e.g. 6.5, 6.6, 6.7, 6.8, 6.9) will round up and those with .4 or less (e.g. 7.0, 7.1, 7.2, 7.3, 7.4) will all round down to the nearest 10. (So all the decimals in brackets here round to 7.)

TASK 1
What does it mean?
Understanding the concept of decimals

Make a list of places where you might see decimals and what they might mean. For example, decimals are used in the price of fuel at a service station – $1.72 per litre.

TASK 2

Kitchen decimals

Identifying decimals in food packaging

You will need:

- box of breakfast cereal or similar product

Discuss with students the nutritional information from a box of breakfast cereal or a similar product. This information tells you the energy, protein, fat, carbohydrate, sodium (salt), sugar contents and much more in a standard serve, and many of these measures are given as decimals.

Nutrition Information (AVERAGE)

Servings per package: 6
Serving size: 35 g (1 metric cup†)

	quantity per serving	% daily intake▲ per serving	per serve with 1/2 cup skim milk	quantity per 100 g
ENERGY	570 kJ	7%	760 kJ	1620 kJ
PROTEIN	2.8 g	6%	7.4 g	7.9 g
FAT, TOTAL	0.6 g	0.9%	0.8 g	1.8 g
- SATURATED	0.1 g	0.4%	0.2 g	0.3 g
CARBOHYDRATE	28.6 g	9%	35.1 g	81.8 g
- SUGARS	3.1 g	3%	9.6 g	8.9 g
DIETARY FIBRE	1.3 g	4%	1.3 g	3.8 g
SODIUM	169 mg	7%	226 mg	485 mg
		%RDI*		
RIBOFLAVIN (VIT B2)	0.42 mg	25%	0.68 mg	1.21 mg
NIACIN	2.5 mg	25%	2.6 mg	7.1 mg
VITAMIN B6	0.4 mg	25%	0.4 mg	1.1 mg
FOLATE	50 µg	25%	56 µg	142 µg
IRON	3.0 mg	25%	3.1 mg	8.6 mg
ZINC	1.8 mg	15%	2.3 mg	5.1 mg

† Cup measurement is approximate and is only to be used as a guide. If you have any specific dietary requirements please weigh your serving.

▲ Percentage daily intakes are based on an average adult diet of 8700 kJ.

* Percentage Recommended Dietary Intake (Aust/NZ).

Ingredients

Corn (90%), sugar, salt, barley malt extract, vitamins (vitamin E, niacin, vitamin B6, riboflavin, folate), minerals (iron, zinc oxide).

**CONTAINS CEREALS CONTAINING GLUTEN.
MAY CONTAIN TRACES OF PEANUTS AND/OR TREE NUTS.**

Choose one healthy item and one unhealthy item from your pantry or fridge. Compare the ingredients and record what you notice about the nutritional information per serve.

TASK 3

Sock ball challenge

Connecting decimals and time

You will need:

* pair of socks

For this task, ask students to roll a pair of socks into a ball and work in pairs.

Stand about 1 m away from a wall. Throw the sock ball against the wall and catch it again ten times while your partner times this with a stopwatch. If you drop the sock ball, you must start again but keep the stopwatch running. Try this several times and for each attempt, record your time to the nearest hundredth of a second, then swap roles.

What was your best time? What was your opponent's best time? Who was the winner?

TASK 4

Let's play

Playing decimal games

You will need:

* calculator
* playing cards
* BLM 34

Explain to students that we can use the place value chart and our knowledge of how decimals work to play lots of great decimal games, using a calculator or cards.

Hundreds	Tens	Ones	.	Tenths	Hundredths	Thousandths
100s	10s	1s		0.1	0.01	0.001

Remember, every time we move to the right on the chart, the numbers we deal with get 10 times smaller.

Up to 100

Play in pairs.

Take turns to add numbers less than 20. First to 100 wins.

(Clue: You can use decimals!)

Wipe-out

Type in a number that uses both whole numbers and decimals. Write the number down.

Now pass the calculator to a partner and say which one of the digits you want to be replaced by a zero. Your opponent can only use subtraction to do this.

The game ends when your opponent has zero on their screen. Now reverse the roles so you have a go at your opponent's number. Make the numbers harder and harder as you go.

Target 100

Play in pairs.

Choose a target range, say 98 to 102.

The first player types in a number less than 10.

Now, taking turns, using only multiplication and numbers less than 10, see who is the first to make a number that sits within the target.

Decimal card challenge

Get 10 playing cards numbered 1 to 10. The 10 will stand for zero in this game. Play with a partner.

Shuffle the cards and deal your opponent 2 cards.

Their task is to make as big a decimal as possible with those 2 cards. (A 4 and 7 will make 0.74 for example).

Reverse the roles to see who has made the larger decimal. 1 point for the winner. Now do the same with 3 cards, then 4, then 5 then finally 6 cards.

Now start the game again but this time the aim is to make the smallest decimal possible.

Biggest wins

Get 10 playing cards numbered 1 to 10. The 10 will stand for zero in this game. Play with a partner.

Write down 2 blank spaces, a decimal point and 2 decimal blank spaces after the decimal point. Your aim is to make the biggest possible number.

Shuffle the cards and turn 1 over. The 2 players must place this digit somewhere in the 4 blank spaces. Now show your opponent where you put this digit. Return the card to the pile of 10 cards, shuffle and select another card. Keep doing this until all 4 blank spaces are filled. The player with the higher number at the end wins the game.

Now try the game with 6 blank spaces, then 8 and finally 10 blank spaces.

OXFORD UNIVERSITY PRESS

TASK 5

Amazing human records

Exploring records that use decimals

You will need:

- tape measure
- stopwatch
- measuring jug
- bucket
- water

Many amazing records have been set throughout history. Use a tape measure and a stopwatch to experience some of these records:

1. **World's tallest human**

   Robert Pershing Wadlow (1918–1940) was once the world's tallest man, standing at 2.72 m. Measure your height and show it as a decimal. How much taller than you was Robert Wadlow?

2. **World's long jump record**

   In 1991 US athlete Mike Powell jumped 8.95 metres setting a new world record.

   Use a tape measure to measure out 8.95 m.

   Try to do a long jump yourself and measure your performance. How far short of Mike Powell's jump was yours?

3. **World's high jump record**

   In 1993 Cuban athlete Javier Sotomayor cleared a height of 2.45 m.

   Use a tape measure to see how high this is. Try to jump up and touch this height.

   How far short of 2.45 m can you jump?

4. **World's 100 m sprint record**

   In 2009 Usain Bolt ran 100 m in 9.58 seconds.

   Get someone to time you running for 9.58 seconds. How far did you run?

   How far from 100 m did you cover?

5. **World shot-put record**

   In 2021 Ryan Crouser of the US broke the world's record for the shot-put with a throw of 23.37 m. The shot weighs 7.26 kg.

   Get a measuring jug and bucket and fill the bucket with 7.26 L of water. Now lift the bucket.

   Now measure out 23.37 m. Imagine throwing the bucket of water that far!

TASK 6

Decimal bingo

Playing bingo with decimals

You will need:

- BLM 35
- cards numbered 1 to 10

How to play:

1. Fill in all 36 blanks on the top bingo board with 2-digit decimals ranging anywhere from 0.00 to 0.99.

2. Ask an opponent to do the same thing on the bottom board with their own 36 2-digit decimals.

3. Now get 10 playing cards numbered from 1 to 10, with the 10 standing for zero.

4. Select 1 (tenths) and put it back into the pile, shuffle and select another card (hundredths). This will form a 2-digit decimal like 0.03 or 0.58. If you have this decimal or one that is one hundredth bigger or smaller, you cross it out. For example, a draw of 0.76 will also include 0.75 and 0.77.

5. The first bingo winner will fill in a line that can be horizontal, vertical or diagonal. The second winner will fill in all 36 decimals.

B	I	N	G	O	!

TASK 7

Batting averages

Applying decimals to cricket

You will need:

- BLM 36

Explain to students that cricket is an example of using decimals in sport. A cricketer's batting average combines both their highest scores and their lowest scores to give an indication of their overall career standard. The batting average is the total number of runs that they have scored in all the test matches they have batted, divided by the number of times they have been dismissed (the times that they were out).

This table shows batting statistics for some of the greatest test cricketers of all time. Use a calculator to find each player's average, then round it to two decimal places.

Player	Career	Test runs	Times out	Average
Donald Bradman	1928–1948	6996	70	
Jan Brittin	1979–1998	1935	22	
Mominul Haque	2013– (still playing)	2860	70	

OXFORD UNIVERSITY PRESS

Player	Career	Test runs	Times out	Average
Herbert Sutcliffe	1924–1935	4555	75	
Vinod Kambli	1993–1995	1084	20	
Kane Williamson	2010– (still playing)	6476	127	
Javed Miandad	1976–1993	8832	168	
Ellyse Perry	2007– (still playing)	752	10	
Graeme Pollock	1963–1970	2256	37	
Kumar Sangakkara	2000–2015	12400	216	
George Headley	1930–1954	2190	36	

TASK 8

Pi challenge

Rounding decimals

You will need:

• BLM 37

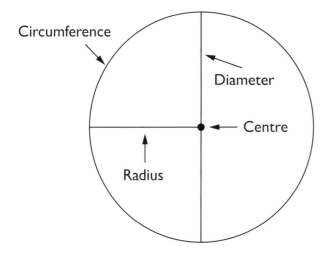

Explain to students that the irrational number Pi is a great example of dealing with rounded decimals. Pi represents how many times the diameter of a circle will divide into its circumference. It is always the same answer, no matter how big or small the circle is. But more amazing is the fact that this number goes on forever and has no pattern to it. Pi has been calculated to trillions of decimal places, so whenever you write down Pi, no matter how many decimal places you use in the answer, it will always be rounded.

Pi = 3.141592653589793 …

Use your knowledge of rounding decimals to round Pi to the number of decimal places in this table.

Round Pi to nearest	Rounded answer
Whole number	
1 decimal place	
2 decimal places	
3 decimal places	
4 decimal places	
5 decimal places	
6 decimal places	
7 decimal places	
8 decimal places	
9 decimal places	
10 decimal places	
11 decimal places	
12 decimal places	
13 decimal places	
14 decimal places	

TASK 9

Problems to solve

Applying decimal skills to unfamiliar situations

1. The radioactive element Demium has a half-life of one day. This means that when it is created in a laboratory, it will have a strength of 1.0. Twenty-four hours later it will have a strength of 0.5. As a decimal, what will the strength of Demium be after five days? As a decimal, what will the strength of Demium be after 10 days?

2. Use the digits 1, 2, 3, 4, 5, 6 and 7 to form an addition sum that equals 3.817 and uses only three addends.

3. Find three consecutive decimals that add up to 1.23.

4. What number am I?

   I am a whole number and a 3-digit decimal. All my digits are different.

   The sum of my four digits is 19.

   When rounded to the nearest whole number I equal 6. Three of my digits are odd.

   My biggest digit is in the hundredths place. What three possible numbers could I be?

5. 0.1 + 23 … This is the start of an equation that uses all the digits from 0 to 9, in numerical order, to equal 100. Can you finish it?

OXFORD UNIVERSITY PRESS

ANSWERS

Task 1 What does it mean?

Answers will vary but may include:

- a thermometer to show body temperature or weather temperatures
- a stopwatch to show sporting performances or challenges
- competition judges to show performance rankings
- on graphs and in statistics to show comparisons or trends
- in money separating dollars from cents
- in division showing remainders
- in cricket, basketball, baseball and many other sports to show performance or winning averages
- on ingredients and food products to show dietary information
- in elections to show party performances
- in medicine to show levels in blood tests.

Task 2 Kitchen decimals

Answers will vary but it is highly likely that the sodium, sugar and fat contents of the less healthy alternative will be much higher.

Task 3 Sock ball challenge

Answers will vary.

Task 4 Let's play

Answers will vary.

Task 5 Amazing human records

Answers will vary.

Task 6 Decimal bingo

Answers will vary.

Task 7 Batting averages

Bradman 99.94. Brittin 87.95. Haque 40.86. Sutcliffe 60.73. Kambli 54.20. Williamson 50.99. Miandad 52.57. Perry 75.2. Pollock 60.97. Sangakkara 57.41. Headley 60.83

Task 8 Pi challenge

Round to the nearest	Rounded answer
Whole number	3
1 decimal place	3.1
2 decimal places	3.14
3 decimal places	3.142
4 decimal places	3.1416
5 decimal places	3.14159
6 decimal places	3.141593
7 decimal places	3.1415927
8 decimal places	3.14159265
9 decimal places	3.141592654
10 decimal places	3.1415926536
11 decimal places	3.14159265359
12 decimal places	3.141592653590
13 decimal places	3.1415926535898
14 decimal places	3.14159265358979

Task 9 Problems to solve

1. 1 day: 0.5
   2 days: 0.25
   3 days: 0.125
   4 days: 0.0625
   5 days: 0.03125
   6 days: 0.015625
   7 days: 0.0078125
   8 days: 0.00390625
   9 days: 0.001953125
   10 days: 0.0009765625.
2. 3.16 + 0.4 + 0.257.
3. 0.40 + 0.41 + 0.42.
4. 6.175 or 6.193 or 6.391.
5. 0.1 + 23 − 4 + 5 + 67 + 8.9 = 100.

UNIT 9 – 100 PER CENT
Percentages

Because percentages can also be expressed as fractions, they too come under the mathematical heading of rational numbers. Percentages present fractions as parts of 100. The term 'percent' comes from two Latin terms: 'per' meaning 'of' and 'centum' meaning 100. Even the symbol for percentage (%) is a variation on the number 100 – it uses a 1 and 2 zeros.

Most problems that deal with percentages involve the conversion of a percentage into its equivalent fractional form and so the more percentage/fraction equivalences that a student can learn to recognise as basic facts, the easier percentage work will be for them. Keep referring back to first principles with your students, relating $\frac{1}{4}$ to one-quarter of 100 (25%) and $\frac{3}{10}$ to $\frac{1}{10}$ of a 100 × 3 (30) and so on.

Brainstorm examples of percentages in sporting contexts, sales and discounts, interest rates, food packaging, test results and so on.

TASK 1
What does it mean?
Understanding the concept of percentage

Discuss examples of when and where we use percentages, such as in sport, on food packaging, in sales, for test results and so on.

List as many different examples of using percentages in everyday life as you can.

TASK 2
Fractions and percentages
Turning a fraction into a percentage
You will need:

- BLM 38
- calculator
- playing cards

Remind students that percentages are really fractions with a denominator of 100; for example, 25% is the same as $\frac{25}{100}$. To turn a fraction into a percentage, all you need to do is multiply it by 100 and then add the % symbol. Students will know many common percentages such as one-half (50%) and one-quarter (25%).

Let's play a game that will give you the chance to see how many fractions/percentage equivalents you know. If you don't know them, use a calculator to find out. For example, to turn $\frac{7}{8}$ into a percentage with the calculator, simply press [7] [÷] [8] [=] *, then multiply this answer by 100. It will give you 87.5. Then simply add the % symbol.*

Get students a set of cards numbered 1 (ace) to 10. After shuffling the cards, ask them to turn two cards over to make a fraction. The smaller of the two numbers will be the numerator and the larger will be the denominator. Then turn this fraction into a percentage.

1. There are 45 possible fractions that you can make in this game, but you only need to make 20.

FRACTIONS AND PERCENTAGES
BLM 38

Question	Numerator	Denominator	Fraction	Percentage	Question	Numerator	Denominator	Fraction	Percentage
1					2				
3					4				
5					6				
7					8				
9					10				
11					12				
13					14				
15					16				
17					18				
19					20				

OXFORD UNIVERSITY PRESS

2. This time use the larger number as the numerator and the smaller number as the denominator. This will produce a number bigger than 1 and a percentage bigger than 100.

FRACTIONS AND PERCENTAGES   BLM 38

Question	Numerator	Denominator	Fraction	Percentage	Question	Numerator	Denominator	Fraction	Percentage
1					2				
3					4				
5					6				
7					8				
9					10				
11					12				
13					14				
15					16				
17					18				
19					20				

Making Maths Count © Peter Maher 2022. This sheet may be photocopied for non-commercial classroom use.

TASK 3

Decimals and percentages

Turning a decimal into a percentage

You will need:

- BLM 39
- dice

Remind students that in the past, decimals were called decimal fractions because decimals, like percentages, are just another way of showing fractions. We convert both decimals and fractions into percentages the same way – simply multiply by 100. When any number is multiplied by 100, you simply move the digits two places to the left on a place value chart. (Explain to students that although moving the decimal point two places to the right will give you the same answer, it is impossible to move the decimal point because it always stays between the ones and the tenths place in any number.)

1. Roll a 6-sided die twice to create a 2-digit decimal (e.g. rolling 4 then 1 will create 0.41) Then convert this decimal into a percentage. The first one has been done for you.

2 rolls	Decimal	Percentage	2 rolls	Decimal	Percentage	2 rolls	Decimal	Percentage
4, 6	0.46	46%						

2 rolls	Decimal	Percentage	2 rolls	Decimal	Percentage	2 rolls	Decimal	Percentage

2. Roll a 6-sided die three times to create a 3-digit decimal (e.g. rolling 3, 2 then 1 will create 0.321). Then convert this decimal into a percentage. The first one has been done for you.

3 rolls	Decimal	Percentage	3 rolls	Decimal	Percentage	3 rolls	Decimal	Percentage
1, 1, 6	0.116	11.6%						

Ask students how we could change a percentage back into a decimal. Explain that we can divide the percentage by 100, because division is the opposite operation to multiplication. For example, divide 12% by 100 to get 0.12. If, for example, the percentage is a mixed number, like $12\frac{1}{2}$%, firstly turn this into a decimal (12.5) and then divide by 100, giving 0.125.

3. Divide by 100 to change these percentages into decimals.

Percentage	Decimal	Percentage	Decimal	Percentage	Decimal
20%		40%		70%	
100%		23%		47%	
88%		92%		133%	
167%		198%		233%	
32.6%		14.7%		28.88%	
156.44%		$6\frac{1}{4}$%		$7\frac{5}{8}$%	
$5\frac{3}{4}$%		$\frac{1}{3}$%			

TASK 4

Test scores

Finding percentages of an amount

You will need:

- BLM 40
- cards numbered 1 to 10
- calculator

OXFORD UNIVERSITY PRESS

Explain to students that when two amounts of the same item are compared, one is often described as a percentage of the other. For example, if 20 apples in a box of 200 apples are bruised and withdrawn from sale, we can say that 20 of the 200 or 10% of the apples were rejected. This means that $\frac{180}{200}$ of the apples, or 90% of the apples, were acceptable. The addition of the two percentages will always equal 100%.

Give the students a set of cards numbered 1 to 10 and ask them to choose pairs of numbers.

1. Calculate what percentage of the larger number selected is the smaller number selected as a fraction, and what percentage of the larger number selected is left. For example, if you select 8 and 2, then 2 represents $\frac{2}{8}$ of 8. Divide 2 by 8 and multiply by 100 to find the percentage: 25%. This leaves $\frac{6}{8}$, so divide 5 by 8 and multiply by 100 to find 75% remaining. Use your knowledge of number facts or a pen and paper to record the answers. Check these answers on a calculator when you have finished.

Numbers selected	Fraction	Percentage	Fraction left	Percentage
3 and 8	$\frac{3}{8}$	37.5%	$\frac{5}{8}$	62.5%

2. Test results are often expressed as percentages. The marks scored can be shown as a fraction: for example, 30 out of 80 is $\frac{30}{80}$, then 30 divided by 80 and multiplied by 100 gives the percentage. Complete the scores in this table.

Test out of	Score	Percentage	Score	Percentage	Score	Percentage	Score	Percentage
80	30		50		70		77	
50	24		37		48		49.5	
75	21		42		60		72	
40	22		25		30		38	
30	12		15		24		27	
25	7		12.5		20		24	
90	10		30		50		80	

TASK 5

AFL percentages

Using percentages in sport

You will need:

- BLM 41
- cards numbered 1 to 20

Explain to students that in AFL football, percentages are used to rank teams on the ladder to separate those that have won a similar number of games. Percentages are calculated by dividing the number of points a team has scored by the number of points scored against them and then multiplying this fraction by 100. The points a team scores and the points scored against them are added up progressively throughout the season. A team with a percentage of about 100 will have scored roughly the same number of points over the season that have been scored against them and are likely to finish about in the middle of the team rankings on the ladder.

Give the students a set of cards numbered 1 to 20 and ask them to complete the AFL ladder for the games listed in the table.

First choose a card to decide the number of goals that a team has scored (each worth 6 points). Next choose a card to decide the number of behinds (a 'behind' is when a player scores between a central and outer post) that a team has scored (each worth 1 point). Then calculate the total number of points for each team. Divide the total points scored by a team by the total points scored against them to find each team's percentage. Then rank each team from 1 to 18 on the ladder, according to its percentage.

Game	Team	Goals	Behinds	Points	Percentage
1.	Adelaide Crows				
	Melbourne				
2.	Essendon				
	Richmond				
3.	Collingwood				
	St Kilda				
4.	Greater Western Sydney				
	Fremantle				
5.	Geelong				
	North Melbourne				
6.	Brisbane Lions				
	Sydney Swans				

Game	Team	Goals	Behinds	Points	Percentage
7.	Hawthorn				
	West Coast Eagles				
8.	Gold Coast Suns				
	Western Bulldogs				
9.	Carlton				
	Port Adelaide				

TASK 6

Tare, net, gross

Relating percentages and weight

You will need:

- BLM 42

Explain to students that when an item is weighed, there are three different types of weights that might be considered. For example, think about a truck carrying rocks or gravel: the tare weight is the weight of the truck when it is empty, the net weight refers to the weight of the rocks and gravel inside the truck, and the gross weight is the combined weight of both the truck and the rocks and gravel inside it. A truck driver needs to pay careful attention to these three weights because the truck has a weight limit for the load it carries, and roads have a weight limit for the truck and its load.

1. Use your knowledge of tare, net and gross weight to complete this table, and work out what percentage of the gross weight of each vehicle is the net weight of its load. (Hint: Subtract the net weight from the gross weight to find the tare weight. Subtract the tare weight from the gross weight to find the net weight. Add the tare weight and the net weight to find the gross weight.)

Vehicle	Load	Tare weight	Net weight	Gross weight	Percentage of gross weight that is net weight
Truck	Furniture	2700 kg	3800 kg		
Car	4 passengers	1250 kg		1500 kg	
Boat	Diesel fuel		400 kg	3800 kg	
Plane	Cargo			80 000 kg	25%

2. Use your knowledge of tare, net and gross weight to complete this table, and work out what percentage of the gross weight of each item is the net weight of its contents. Look in your

fridge or pantry to add some items of your own. (Hint: Subtract the net weight from the gross weight to find the tare weight. Subtract the tare weight from the gross weight to find the net weight. Add the tare weight and the net weight to find the gross weight.)

Item	Tare	Net weight	Gross weight	Percentage of gross weight that is net weight
Can of tuna		95 g	115 g	
Tomato paste	4 g		44 g	
Gravy sachet	5 g	165 g		
Packet of cashews		750 g	753 g	
Oyster sauce	190 g	255 g		
Breakfast cereal		495 g		99%

TASK 7

Sales and discounts

Applying percentages to shopping

You will need:

- BLM 43

Remind students that when products are offered for sale at a discount, we usually see three different pieces of information: the regular price (the price of the item before the sale), the discount (usually shown as a percentage of the regular price) and the sale price (the regular price minus the discount).

1. The clothing store 'Threads R Us' is having their 'Any day ending in a Y' super sale, which lasts all month. Use your knowledge of percentages to complete this table:

Item	Regular price	Discount	Amount saved	Sale price
Jumper	$60	25%		
Jacket	$150		$50	
Cardigan	$40			$25
Shorts		10%	$3	
T-Shirt			$5	$20
Business shirt		30%		$56
Long pants	$70			$40
Rain jacket		60%	$36	
Socks			$4	$2

OXFORD UNIVERSITY PRESS

Supermarkets often have weekly specials to attract regular customers. Sometimes supermarkets show discounts as fractions (e.g. 'half-price') and sometimes as money saved (e.g. 'ten dollars off') and sometimes as percentages (e.g. 25% discount).

2. Find a supermarket catalogue to help you create a table like the one above.

TASK 8

Commission

Understanding percentage payments

You will need:

- BLM 44

Explain to students that many sales staff work 'on commission'. This usually means that as well as a base salary, they receive extra payments according to how much they sell. In some businesses, like real estate agencies, sales staff get paid a percentage of the value of the homes they sell, so the more expensive the home that is sold, the more commission they earn.

1. Use your knowledge of how percentages work to complete this table, assuming the real estate agent earns 2% commission on every house sold.

Suburb	House price	Estate agent's commission in dollars
Broadmeadows	$555 000	
Carindale	$580 000	
Casuarina	$615 000	
Scarborough	$990 000	
Gold Coast	$1 000 000	
Battery Point	$1 200 000	
North Adelaide	$1 340 000	
Byron Bay	$1 500 000	
South Yarra	$1 700 000	
Elizabeth Bay	$2 050 000	
Pymble	$2 100 000	
Toorak	$4 400 000	

2. Many sales staff earn a base salary and then a commission as well. Jack sells shoes at his local 'Heel N' Sole' shoe shop. He gets paid $300 per week plus 1% of what he sells. Below are Jack's weekly shoe sales for the Christmas period. Can you complete the table?

Week	Shoe sales	Commission	Base wage	Take home pay
December 1–7	$13 450		$300	
December 8–14		$186	$300	

Week	Shoe sales	Commission	Base wage	Take home pay
December 15–21			$300	$540
December 22–28			$300	$680

TASK 9

Keep it simple

Understanding simple interest rates

You will need:

- BLM 45

Explain to students that whenever you deposit money into a bank or a building society, you earn interest on that money. The interest is paid into your account monthly, quarterly or yearly.

On the other hand, if you loan money from a bank or a building society, say if you want to buy a car or a house, then you will pay interest on that money.

Interest rates are shown as percentages. If, for example you borrow $10 000 at an interest rate of 5% and aim to pay the bank back in four years, you will pay $10 000 + 5% × 4 (20% or $2 000). So you will have paid the bank $12 000 over a four-year loan period.

1. Can you complete this deposit table? The first one has been done for you.

Deposit	Interest rate	Term of deposit	Interest paid yearly	Total interest paid	Final deposit amount
$1000	2%	2 years	$20	$40	$1040
$5000	3%	3 years			
$10 000		5 years			$11 000
$20 000	4%		$800		$26 400
$100 000	2.5%				$110 000

2. Congratulations – you have just bought a new home! Can you calculate how much money you will need to pay back the bank on these loan amounts? The first one has been done for you.

Money borrowed	Length of loan	Interest rate	Loan + Interest
$100 000	10 years	5%	$100 000 + 10 × 5% (50% = $50 000) = $150 000
$300 000	12 years	6%	
$500 000	20 years	6.5%	
$700 000	20 years	7%	
$900 000	25 years	8%	
$1 000 000	30 years	10%	

OXFORD UNIVERSITY PRESS

TASK 10

Problems to solve

Applying percentage skills to unfamiliar situations

1. After spending 95% of her money at the Royal Easter Show, Claire had $8 left. How much money did Claire take to the Royal Easter Show?

2. A real estate agent charged a commission of $10 000 for the sale of a house, plus 6% for every dollar over $1 000 000 that the house was sold for. The agent earned $14 200 for the sale of the house. How much did the house sell for?

3. After completing 60% of a software update, a smartphone had 14 minutes to go before the installation was complete. How long did the update take altogether?

4. After Round 2 of the AFL season, the Essendon's percentage was 80. They had scored a total of 100 points. How many points must have been scored against them?

5. After completing two maths tests, a student has an average of 85.5%. Each test was marked out of 100. What is the lowest possible score that the student could have received on one of the tests?

ANSWERS

Task 1 What does it mean?

Answers will vary. The table shows some examples:

Where?	When?
Football	To calculate a team's percentage (Points For/Points Against × 100)
Food packages	To show salt, sugar, protein, carbohydrate levels
Tests	To change a score to a mark out of 100 to show rankings
Businesses	To show profits or losses over time and sales growth rates
Banks	When applying interest rates to loans and account savings
Surveys	To see patterns and trends after data collection
Drinks	To show sugar or alcohol content
Employment	To see how many people are in or out of work or working part time or full time
Climate	To view rainfall in a month or year compared to averages, to see the effects of climate change on temperatures and sea levels
Population	To see growth rates or declines, to compare the sizes of various groups of people in the community

Task 2 Fractions and percentages

1. The 45 possibilities are:

$\frac{1}{2}$ 50%, $\frac{1}{3}$ 33.33%, $\frac{1}{4}$ 25%, $\frac{1}{5}$ 20%, $\frac{1}{6}$ 16.67%, $\frac{1}{7}$ 14.29%, $\frac{1}{8}$ 12.5%, $\frac{1}{9}$ 11.11%, $\frac{1}{10}$ 10%,

$\frac{2}{3}$ 66.67%, $\frac{2}{4}$ 50%, $\frac{2}{5}$ 40%, $\frac{2}{6}$ 33.33%, $\frac{2}{7}$ 28.57%, $\frac{2}{8}$ 25%, $\frac{2}{9}$ 22.22%, $\frac{2}{10}$ 20%,

$\frac{3}{4}$ 75%, $\frac{3}{5}$ 60%, $\frac{3}{6}$ 50%, $\frac{3}{7}$ 42.86%, $\frac{3}{8}$ 37.5%, $\frac{3}{9}$ 33.33%, $\frac{3}{10}$ 30%,

$\frac{4}{5}$ 80%, $\frac{4}{6}$ 66.67%, $\frac{4}{7}$ 57.14%, $\frac{4}{8}$ 50%, $\frac{4}{9}$ 44.44%, $\frac{4}{10}$ 40%,

$\frac{5}{6}$ 83.33%, $\frac{5}{7}$ 71.43%, $\frac{5}{8}$ 62.5%, $\frac{5}{9}$ 55.56%, $\frac{5}{10}$ 50%, $\frac{6}{7}$ 85.71%, $\frac{6}{8}$ 75%, $\frac{6}{9}$ 66.67%, $\frac{6}{10}$ 60%,

$\frac{7}{8}$ 87.5%, $\frac{7}{9}$ 77.78%, $\frac{7}{10}$ 70%, $\frac{8}{9}$ 88.89%, $\frac{8}{10}$ 80%, $\frac{9}{10}$ 90%.

2. The 45 possibilities are:

$\frac{2}{1}$ 200%, $\frac{3}{1}$ 300%, $\frac{3}{2}$ 150%, $\frac{4}{1}$ 400%, $\frac{4}{2}$ 200%, $\frac{4}{3}$ 133.33%, $\frac{5}{1}$ 500%, $\frac{5}{2}$ 250%, $\frac{5}{3}$ 166.67%,

$\frac{5}{4}$ 125%, $\frac{6}{1}$ 600%, $\frac{6}{2}$ 300%, $\frac{6}{3}$ 200%, $\frac{6}{4}$ 150%, $\frac{6}{5}$ 120%, $\frac{7}{1}$ 700%, $\frac{7}{2}$ 350%, $\frac{7}{3}$ 233.33%, $\frac{7}{4}$ 175%,

$\frac{7}{5}$ 140%, $\frac{7}{6}$ 116.67%, $\frac{8}{1}$ 800%, $\frac{8}{2}$ 400%, $\frac{8}{3}$ 266.67%, $\frac{8}{4}$ 200%, $\frac{8}{5}$ 160%, $\frac{8}{6}$ 133.33%, $\frac{8}{7}$ 114. 29%,

$\frac{9}{1}$ 900%, $\frac{9}{2}$ 450%, $\frac{9}{3}$ 300%, $\frac{9}{4}$ 225%, $\frac{9}{5}$ 180%, $\frac{9}{6}$ 150%, $\frac{9}{7}$ 128.57%, $\frac{9}{8}$ 112.5%, $\frac{10}{1}$ 1,000%,

$\frac{10}{2}$ 500%, $\frac{10}{3}$ 333.33%, $\frac{10}{4}$ 250%, $\frac{10}{5}$ 200%, $\frac{10}{6}$ 166.67%, $\frac{10}{7}$ 142.86%, $\frac{10}{8}$ 125%, $\frac{10}{9}$ 111.11%.

Task 3 Decimals and percentages

1. Answers will vary.

2. Answers will vary.

3.

Percentage	Decimal	Percentage	Decimal	Percentage	Decimal	Percentage	Decimal
20%	0.2	40%	0.4	70%	0.7	100%	1.0
23%	0.23	47%	0.47	88%	0.88	92%	0.92
133%	1.33	167%	1.67	198%	1.98	233%	2.33
32.6%	0.326	14.7%	0.147	28.88%	0.2888	156.44%	1.5644
$6\frac{1}{4}$%	0.625	$7\frac{5}{8}$%	0.7625	$5\frac{3}{4}$%	0.0575	$\frac{1}{3}$%	0.0033

Task 4 Test scores

1. The 45 possibilities are:

1 and 2 50% and 50%, 1 and 3 33.33% and 66.67%, 1 and 4 25% and 75%, 1 and 5 20% and 80%, 1 and 6 16.67% and 83.33%, 1 and 7 14.29% and 85.71%, 1 and 8 12.5% and 87.5%, 1 and 9 11.11% and 88.89%, 1 and 10 10% and 90%, 2 and 3 66.67% and 33.33%, 2 and 4 50% and 50%, 2 and 5 40% and 60%, 2 and 6 33.33% and 66.67%, 2 and 7 28.57% and 71.43%, 2 and 8 25% and 75%, 2 and 9 22.22% and 77.78%, 2 and 10 20% and 80%, 3 and 4 75% and 25%, 3 and 5 60% and 40%, 3 and 6 50% and 50%, 3 and 7 42.86% and 57.14%,

3 and 8 37.5% and 62.5%, 3 and 9 33.33% and 66.67%, 3 and 10 30% and 70%, 4 and 5 80% and 20%, 4 and 6 66.67% and 33.33%, 4 and 7 57.14% and 42.86%, 4 and 8 50% and 50%, 4 and 9 44.44% and 55.56%, 4 and 10 40% and 60%, 5 and 6 83.33% and 16.67%, 5 and 7 71.43% and 28.57%, 5 and 8 62.5% and 37.5%, 5 and 9 55.56% and 44.44%, 5 and 10 50% and 50%, 6 and 7 85.71% and 14.29%, 6 and 8 75% and 25%, 6 and 9 66.67% and 33.33%, 6 and 10 60% and 40%, 7 and 8 87.5% and 12.5%, 7 and 9 77.78% and 22.22%, 7 and 10 70% and 30%, 8 and 9 88.89% and 11.11%, 8 and 10 80% and 20%, 9 and 10 90% and 10%.

2.

Test out of	Score	Percentage	Score	Percentage	Score	Percentage	Score	Percentage
80	30	37.5%	50	62.5%	70	87.5%	77	96.25%
50	24	48%	37	74%	48	96%	49.5	99%
75	21	28%	42	56%	60	80%	72	96%
40	22	55%	25	62.5%	30	75%	38	95%
30	12	40%	15	50%	24	80%	27	90%
25	7	28%	12.5	50%	20	80%	24	96%
90	10	11.11%	30	33.33%	50	55.56%	80	88.89%

Task 5 AFL percentages

Answers will vary.

Task 6 Tare, net, gross

1.

Vehicle	Items	Tare weight	Net weight	Gross weight	Percentage of gross weight that is net weight
Truck	Furniture	2700 kg	3800 kg	6500 kg	58.46%
Car	4 passengers	1250 kg	250 kg	1500 kg	16.67%
Boat	Diesel fuel	3400 kg	400 kg	3800 kg	10.53%
Plane	Cargo	60 000 kg	20 000 kg	80 000 kg	25%

2.

Item	Tare	Net weight	Gross weight	Percentage of gross weight that is net weight
Can of tuna	20 g	95 g	115 g	82.61%
Tomato paste	4 g	40 g	44 g	90.91%
Gravy sachet	5 g	165 g	170 g	97.06%
Packet of cashews	3 g	750 g	753 g	99.60%
Oyster sauce	190 g	255 g	445 g	57.30%
Breakfast cereal	5 g	495 g	500 g	99%

Task 7 Sales and discounts

1.

Item	Regular price	Discount	Amount saved	Sale price
Jumper	$60	25%	$15	$45
Jacket	$150	33.33%	$50	$100
Cardigan	$40	37.5%	$15	$25
Shorts	$30	10%	$3	$27
T-Shirt	$25	20%	$5	$20
Business shirt	$80	30%	$24	$56
Long pants	$70	42.86%	$30	$40
Rain jacket	$60	60%	$36	$24
Socks	$6	66.67%	$4	$2

2. Answers will vary.

Task 8 Commission

1.

Suburb	Median house price	Commission in dollars
Broadmeadows	$555 000	$11 100
Carindale	$580 000	$11 600
Casuarina	$615 000	$12 300
Scarborough	$990 000	$19 800
Gold Coast	$1 000 000	$20 000
Battery Point	$1 200 000	$24 000
North Adelaide	$1 340 000	$26 800
Byron Bay	$1 500 000	$30 000
South Yarra	$1 700 000	$34 000
Elizabeth Bay	$2 050 000	$41 000
Pymble	$2 100 000	$42 000
Toorak	$4 400 000	$88 000

OXFORD UNIVERSITY PRESS

2.

Week	Shoe sales	Commission	Base wage	Take home pay
December 1–7	$13450	$134.50	$300	$434.50
December 8–14	$18600	$186	$300	$486
December 15–21	$24000	$240	$300	$540
December 22–28	$38000	$380	$300	$680

Task 9 Keep it simple

1.

Deposit	Interest rate	Term of deposit	Interest paid yearly	Total interest paid	Final deposit amount
$1000	2%	2 years	$20	$40	$1040
$5000	3%	3 years	$150	$450	$5450
$10000	2%	5 years	$200	$1000	$11000
$20000	4%	8 years	$800	$6400	$26400
$100000	2.5%	4 years	$2500	$10000	$110000

2.

Money borrowed	Length of loan	Interest rate	Loan + Interest
$100000	10 years	5%	$100000 + 10 × 5% (50% = $50000) = $150000
$300000	12 years	6%	$300000 + 12 × 6% (72% = $216,000) = $516000
$500000	20 years	6.5%	$500000 + 20 × 6.5% (130% = $650000) = $1,150,000
$700000	20 years	7%	$700000 + 20 × 7% (140% = $980000) = $1680000
$900000	25 years	8%	$900000 + 25 × 8% (200% = $1800000) = $2,700,000
$1000000	30 years	10%	$1000000 + 30 × 10% (300% = $3000000) = $4000000

Task 10 Problems to solve

1. $8 must represent 5% of the original amount of money Claire took to the show. 5% is $\frac{1}{20}$. $8 × 20 = $160.
2. $4200 represents 6% of $70000 so the house must have sold for $1070000.
3. 14 minutes must represent 40% of the time. Thus 10% must be 3.5 minutes. So the update must take 35 minutes to install.
4. If 100 points represents 80% of a total, that total must be $100 \times \frac{100}{80} = 125$ points.
5. 85.5% after 2 tests means that the student must have scored 171 out of 200 marks. The lowest possible total for 1 of the tests is 71, assuming that the other was 100.

UNIT 10 – DO YOU MEASURE UP?
Length, perimeter and area

Length, perimeter and area can be measured both formally and informally. Informal measurement can use all sorts of things as measuring units, providing the things are the same size, like spoons or pieces of paper. Formal measurement uses units such as metres or square centimetres.

The metric system of measurement was devised in the late 18th century in France and was based on the geographical structure of the Earth. The metre was one ten-millionth of the distance from the North Pole, through France, to the Equator. Based on the number 10, the system was far more logical than the imperial system, which was used in Australia until the 1970s. The imperial system of measurement uses units of length such as inch, foot, yard and mile.

After using informal units to measure length and perimeter, start using the formal units of the metre and centimetre with a heavy emphasis on estimation prior to measurement. Most of the measuring we do in our everyday lives involves estimation, which, in turn, helps to crystallise in the mind of the student the measurement unit under review.

Tell students that the term 'perimeter' comes from two Greek words 'peri', meaning 'around', and 'metro' meaning 'to measure', so perimeter means 'to measure around' something. Area is a far more sophisticated concept due to the introduction of a second dimension and is measured in 'square' units. Allow students to cover spaces within boundaries with various items and allow them to discover the formula of 'length × width' in their investigations. In brainstorming sessions, think about lengths and heights involved in sporting activities, the heights of the students themselves, as well as the measurements of furniture and rooms within the school and at home.

TASK 1
What does it mean?
Understanding the concept of measurement

Why do you think we measure things? What sorts of things do we measure and why?

How can we measure things? What kinds of tools do we use?

TASK 2

Table it

Measuring length with informal units

You will need:

- coloured pencils
- measuring unit (e.g. paperclips)

Remind students that length can be measured in many different ways.

Ask students to work in groups using coloured pencils (of identical length) to measure the length and width of a desk or table.

Then ask them to find a measuring unit that is smaller than a pencil, such as a paperclip.

Put the paperclip next to a pencil. Which is longer? Estimate how many paperclips it will take to measure the length and width of the table. Then measure the table – how does this compare to the measurements using pencils?

Now find something else in the classroom that you can measure using pencils and paperclips. Draw a picture of what you are measuring and how you measured it.

TASK 3

Pencil perimeter

Finding perimeter with informal units

You will need:

- pencils

Collect some pencils, making sure that they are all the same length, or very close to the same length. You will be doing some work which deals with PERIMETER.
  Find a rectangular surface, like a study table or a breakfast bar.
  Put 1 pencil on the edge of this surface, starting where two corners meet.
  Estimate how many pencils, touching end to end, would be needed to line up to match the length of this surface.
  Now measure the length of the surface in pencils touching end to end.
  Estimate how many pencils, touching end to end, would be needed to line up to match the width of this surface
  Now measure the width of the surface in pencils touching end to end.
  Now count how many pencils it will take to find the perimeter of this surface. (Look for a clever shortcut. Do we need to measure all 4 edges?)
  As you look around your kitchen you will see many rectangular surfaces like the fridge door, a cupboard door or tiles on the floor.
  Estimate and then find the perimeter of these rectangles using pencils.
  Draw yourself finding the perimeter of one of these shapes and write down how you did it.

Encourage students to self-discover the fact that the opposite sides of a rectangle are of equal length. If students would like to reinforce the concept of perimeter, use different materials to pencils, like pieces of A4 paper or building blocks.

TASK 4

Heel and toe

Understanding uniform units of measurement

An interesting way of measuring length is to count in 'feet' by putting one foot in front of another, touching heel to toe. Let's measure the length of some rooms in your house in this way.

1. **Your bedroom**

   Find which wall will be the length of your bedroom.

   Do 5 measurements using your feet, then estimate the number required to cover the length of your room.

   Now continue and measure the length of your bedroom this way.

   Now get an adult to measure the length of your bedroom in this way.

   Why were the 2 measurements different?

2. **The living room**

   Find which wall will be the length of your living room.

   Do 5 measurements using your feet, then estimate the number required to cover the length of this room.

   Now continue and measure the length of the living room this way.

   Now get an adult to measure the length of the living room in this way.

   Why were the 2 measurements different?

3. **Backyard**

   Find a distance in your backyard that you can measure in this way.

   Do 5 measurements using your feet, then estimate the number required to cover the length you have chosen.

   Now continue and measure the length in this way.

   Now get an adult to measure the same length in this way.

   Why were the 2 measurements different?

TASK 5

Screen time

Measuring length with formal units

You will need:

- ruler or measuring tape
- calculator

Ask students to think about all the devices they have at school and at home that have screens (e.g. televisions, tablets, smart phones, computers and gaming consoles). Explain to students that traditionally, screen size is always measured diagonally, not horizontally, and is usually presented in inches. This is because televisions were created originally in Britain and further developed in the USA, and these countries use the imperial system of measurement. If necessary, show students a ruler or tape measure that uses inches to familiarise them with this unit of measurement.

Ask students to find as many screens as they can and to measure the diagonal screen size of each device in centimetres. By dividing the number of centimetres by 2.54, they can then convert into inches. They can use a calculator for this task and round each answer to the nearest inch.

TASK 6

Paper areas

Finding area with informal units

You will need:

- A4 paper
- A3 paper
- sticky tape

Students can use a piece of A4 paper to measure the area of different surfaces around the classroom, school or at home.

They should estimate then record the actual measurement (in terms of the number of pieces of paper) for each item.

Next, ask students to use a piece of A3 paper (or tape two pieces of A4 paper together) and repeat the task for the same items.

What have you noticed about the answers?

Have you discovered a short cut that will help you find the area of these rectangles faster? What is it?

TASK 7

Rank your rooms

Measuring and comparing areas with formal units

You will need:

- A4 paper
- ruler
- tape measure
- calculator

Explain to students that while measuring area with informal units can work, we need to use uniform formal units when we want real accuracy. Area is always measured in square units, like square centimetres and square metres.

Ask students to make a square centimetre piece of paper, using a ruler to help them. Then have them work in groups to measure out a square metre on the floor, using a tape measure if necessary.

Can you work out how many square centimetres make a square metre?

Use the tape measure to find the area of five surfaces either in your house or in your school (use a short cut if you can).
Which surface has the largest area?

Which surface has the smallest area?
If you're at school, go out into the school yard. How many square metres do you estimate there will be in the area of this space?'
If you're at home, go outside and find an open space like a small park, tennis court, soccer pitch. How many square metres do you estimate there will be in the area of this space?
Can you measure this area in square metres? (A calculator will help.)

TASK 8

Fly on the ceiling

Drawing to scale

You will need:

- paper
- pencils

Explain to students that builders use scale drawings made by architects to show them the sizes and positions of the walls, doors, windows and stairs that they need to construct rooms. Ask students to construct a scale drawing of their own room, using one square centimetre and one centimetre to represent one square metre and one metre, respectively

Pretend that you are a fly positioned exactly in the middle of your bedroom ceiling. You look down to see your bedroom floor.

Make a scale drawing of what your bedroom floor will look like from the ceiling. Make sure to measure the sizes of the things that are normally a part of your room. Make sure to write these measurements on the scale drawing.
On your plan you need to include the length and width of your walls, bed, cupboards and furniture within the room.

TASK 9

My measuring system

Creating a new way to measure

You will need:

- paper
- pencils

Explain to students that the imperial system of measurement uses the inch, foot, yard and mile as units of length. They are connected in such a way that 12 inches equal a foot, 3 feet equal a yard and 1760 yards equal a mile. These units were initially based on the human body (an inch was the distance across the thumb.)

Remind students that the metric system of measurement was originally based on the distance from the North Pole to the Equator. When this distance is divided up into 10 million equal pieces, we get the metre. Millimetres, centimetres and kilometres are all connected to the metre.

Your task is to create a new standard unit, based on something in your life. Describe your standard unit and why you chose it. Then break it up into smaller units, giving these units new names. Then create a name for multiple lengths of your unit.

For example, the Bear System, could be based on the length of your favourite soft toy. One bear might be similar to a metre, one hundredth of a 'bear' might be called a 'ted', and 100 bears might be called a 'teddy' and so on.

Draw and label pictures of your standard units, as well as the related smaller and longer units.

TASK 10
Problems to solve
Applying measurement skills to unfamiliar situations

1. A rectangle has a length of x metres and a width of y metres, and $y - x = -9$ and $x + y = 15$. What is the area of this rectangle?

2. A snail can travel 1 m in a minute, a rabbit can hop 100 m in a minute, a car can travel 1 km in a minute and a rocket can travel 100 km in a minute. In one minute, how far can a snail, a rabbit, a car and a rocket travel altogether? Write your answer in metres and kilometres.

3. How many rectangles, each having its own area, can be seen in a tennis court?
   (Clue: There are many more than 6!)

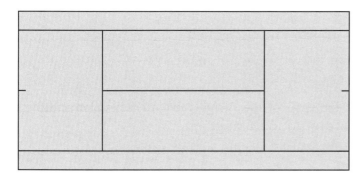

4. A truck leaves a Melbourne warehouse at 6am heading for Sydney on the Hume Highway. The truck travels at an average speed of 60 km/hr. At 11am on the same day, a car leaves the same warehouse heading for Sydney and drives at an average speed of 100 km/hr. How far from Melbourne and at what time will the truck and the car be side by side on the Hume Highway?

ANSWERS

Task 1 What does it mean?

Answers will vary but common responses will include:

Why we measure things: To know accurately the height, length, width, perimeter, area, mass, capacity, volume, angle or temperature of things.

What things we measure: People's height and weight, distances between towns or countries, liquids in a bottle, area of a room that may need painting or carpeting, temperature of a sick person.

How we measure things: Using informal units or formal units like millimetres, centimetres, metres, kilometres, grams, kilograms, millilitres, litres, degrees, square metres or hectares.

Task 2 Table it

Answers will vary.

A paperclip is shorter in length than a pencil so estimates should be greater for the smaller unit.

Task 3 Pencil perimeter

Answers will vary.

Task 4 Heel and toe

Answers will vary.

Ask students why the number of steps required for an adult to cover a length in this way will be fewer. This brings up the need for uniformity in measurement and the need for standard units such as centimetres and metres.

Task 5 Screen time

Answers will vary.

Task 6 Paper areas

Answers will vary.

Students should appreciate that the areas of the kitchen table, the television and the bed should require only half as many pieces of A3 as A4 sheets.

The shortcut to finding the area of rectangles requires length × width.

Task 7 Rank your rooms

There are 10 000 sq cm in 1 sq m.

Answers will vary.

Task 8 Fly on the ceiling

Answers will vary.

Task 9 My measuring system

Answers will vary.

Task 10 Problems to solve

1. The rectangle must be 12 m long and 3 m wide with an area of 36 sq m.
2. 101 101 m or 101.101 km.
3. There are 13 rectangles on the court.
4. At 6:30pm the car and the truck will be 750 km from Melbourne when they will be side by side. When the car leaves the warehouse, the truck will be 300 km away. It will gain 40 km/hr on the truck. 300 divided by 40 = 7.5. In 7.5 hours the car will be 750 km from Melbourne.

UNIT 11 – WEIGHTY MATTERS
Mass and capacity

One of the great strengths of the metric system is the interconnectivity of the concepts of length, volume, mass and capacity. For example, 1 litre of water weighs 1 kilogram and that 1 gram is the weight of 1 millilitre of water. A MAB 1000 block is 10 cm on each edge and has a volume of 1000 cubic centimetres, and this volume is the amount of space occupied by a litre of water, which weighs 1 kilogram. If we now consider one one-thousandth of these units, we can see that one MAB unit is 1 cm on each edge and has a volume of 1 cubic centimetre. This space will hold 1 millilitre of water which will weigh 1 gram. Encourage students to explore connections between measuring units for mass and capacity. When brainstorming, think about using hefting to estimate weights and capacities, introduce scales and measuring jugs, and think about situations where we most commonly measure mass and capacity.

TASK 1
What does it mean?
Understanding the concepts of mass and capacity

1. Can you think of some examples of what and when we might weigh something?
2. Can you think of some examples of what and when we might find the capacity of something?

Discuss examples of when and why we use mass and capacity.

TASK 2
Exploring mass and capacity
Estimating and hefting

You will need:

- five random items from home or classroom
- 2 L bottles of milk or water
- bags of potatoes
- cups
- small saucepans

Explain to students that they can use their measuring skills to see how accurate they can be when dealing with mass and capacity. Remind the students that 'hefting' is estimating measurement by lifting or holding something with your hands.

Choose five items at home or in the classroom and use your hands as scales to compare the mass of these items. Now rank them in order from lightest to heaviest.

Work with a partner – ask them to estimate and record the mass of the same five items. Did they rank them in the same order?

Provide groups of students with a 2 L bottle of milk or water and a bag of potatoes. Ask them to estimate how many potatoes it would take to weigh the same as the liquid.

Provide groups of students with a cup and a small saucepan. Ask them to estimate how many cups of water it would take to fill the saucepan, then ask them to check their estimations by carrying this out.

TASK 3

Kitchen mass and capacity

Using mass and capacity in the kitchen

You will need:

- range of common kitchen items, or supermarket catalogue with images of food and drink
- BLM 46

Provide students with a range of common kitchen items, or supermarket catalogues showing images of food and drink. Alternatively, students could complete this activity at home. Ask them to record items measured by their mass or capacity. They might like to try estimating them with their hands first.

1. Things measured by their mass:

Item	Mass of item

2. Things measured by their capacity:

Item	Capacity of item

OXFORD UNIVERSITY PRESS

TASK 4

Recipe measures

Using mass and capacity in cooking

You will need:

- recipe

Explain to students that when creating a successful recipe, a cook needs to experiment with a combination of ingredients and cooking times to achieve something that tastes good. Recipes contain instructions that use ingredients shown in grams, kilograms, litres and millilitres. Sometimes these amounts are also shown in teaspoons or tablespoons or in cups or fractions of a cup. Explore some recipes with the students.

Find a recipe for something you can cook or bake yourself, taking note of the measurements provided. Write examples of ingredients measured in metric units. Use your research skills. How many grams are there in a teaspoon and in a tablespoon? How many millilitres are there in a cup or half a cup?

TASK 5

Shower time

Measuring water flow

You will need:

- bucket
- stopwatch or smartphone
- measuring jug or cup

Remind students that one of the ways we can help conserve or use water wisely is to limit the length of our showers. Ask them to calculate the amount of water their family uses each day when showering.

1. Get a bucket and a stopwatch (you can find one on a smartphone).
2. Place the bucket directly under a shower head.
3. Turn the taps on full for exactly 1 minute, making sure the water all goes into the bucket.
4. Now get a measuring jug or measuring cup to see how much water came out of the shower in 1 minute. Record the result.
5. Next time you have a shower, use a stopwatch to time it. Multiply the number of minutes by the amount of water you recorded in Step 4, to work out the total amount of water you used.
6. Use your research skills to find the average time or water flow for a water-wise shower. How does your shower compare?

TASK 6

It weighs ... it holds ...

Estimating and measuring mass and capacity

You will need:

- BLM 47

Explain to students that 1 litre of water weighs 1 kilogram and that 1 gram is the weight of 1 millilitre of water.

Find a range of items that are measured in grams. Estimate and then use scales to measure the mass of each item. How accurate were your estimations?

Item	Estimated mass in grams	Actual mass in grams
1		
2		
3		
4		

Fill eight different containers with water to different levels. Estimate and then use a measuring jug to measure the amount of water in each container in millilitres. How accurate were your estimations?

Item	Estimated capacity in millilitres	Measured capacity in millilitres
1		
2		
3		
4		

TASK 7

Which units?

Matching measures to metric units

You will need:

- BLM 48

When we find the mass or the weight of an object, the units used are usually grams or kilograms. When we find the capacity of something, the units used are usually millilitres or litres. See if you can match the mass or capacity of these objects with the most appropriate measurement:

OXFORD UNIVERSITY PRESS

Object	Mass or capacity	Possible answers
Water in a full watering can		80 kg
Water in an Olympic pool		10 kg
Capacity of a juice box		10 000 kg
A city's average rainfall in 1 year		60 L
Weight of a man or woman		50 mL
Weight of a primary student		10 L
Weight of a truck		3 kg
Milk in a bottle		100 g
Weight of a sack of potatoes		30 L
Medicine in an eyedropper		35 kg
A mouthful of water		1500 kg
Water in a spa		650 mm
Weight of a baby		5 g
Weight of a pencil		3000 L
Water in a full bathtub		2 L
Water in a washing machine		2 000 000 L
Weight of a calculator		250 mL
Water in a full kitchen sink		200 L
Weight of a car		5 mL

TASK 8

Problems to solve

Applying mass and capacity skills to unfamiliar situations

You will need:

1. A chemist fills a beaker with 1 L of an experimental liquid. She then removes 500 mL of the liquid and adds 400 mL of water to dilute the mixture. She then removes half of the liquid left in the beaker.

   A. How much mixture is now in the beaker?

   B. How much of the final liquid in the beaker is the original experimental liquid?

2. Rainfall, although a liquid, is measured in millimetres, not millilitres as you would think. This is because it is measured by the distance it travels up a rain gauge. Melbourne's average annual rainfall is 650 mm. In a three-year period, Melbourne received 580 mm, 540 mm and

375 mm of rain. How much rain must fall the next year to bring the four-year total back to average?

3. A carton of 10 cans of beans weighs 8.2 kg. The cardboard carton weighs 300 g and each tin can that the beans come in weighs 250 g. What is the weight of beans in each of the cans?

4. Babies and young children grow very quickly. Statistics show that the average baby weighs about 3.2 kg at birth. They triple this weight by the age of 1. They add on their birth weight again by the time they are 2. They add on half their birth weight again by the time they reach 3 years of age. They add on about the same weight for the next 2 years and average a weight of 18.6 kg by the time they reach the age of 5.

How much weight does the average child weigh at the age of 4?

ANSWERS

Task 1 What does it mean?

Answers will vary but may include:

1. Weight: when we weigh ourselves, when we weigh ingredients for a recipe, when we weigh a suitcase for travelling, when the maximum weight of a vehicle needs to be checked, when we are doing experiments, etc.

2. Capacity: when drinking water, when measuring chlorine needed in a swimming pool, when using a recipe, when working out a water bill, when mixing chemicals in a bucket, when filling the car with petrol or diesel, etc.

Task 2 Exploring mass and capacity

Answers will vary.

Task 3 Kitchen mass and capacity

Answers will vary.

Task 4 Recipe measures

A teaspoon equals 5 g. A tablespoon equals 15 g. A cup equals 250 mL. A half a cup equals 125 mL.

Task 5 Shower time

Water flows from a shower head at about 10 L per minute, so a 5-minute shower will use 50 L of water. Cut that down to 3 minutes and you can save over 7000 L of water in a year.

Task 6 It weighs … it holds …

Answers will vary.

OXFORD UNIVERSITY PRESS

Task 7 Which units?

Object	Mass or capacity
Water in a full watering can	10 L
Water in an Olympic pool	2 000 000 L
Capacity of a box juice	250 mL
A city's average rainfall in 1 year	650 mm
Weight of a man or woman	80 kg
Weight of a primary student	35 kg
Weight of a truck	10 000 kg
Milk in a bottle	2 L
Weight of a sack of potatoes	10 kg
Medicine in an eyedropper	5 mL
A mouthful of water	50 mL
Water in a spa	3000 L
Weight of a baby	3 kg
Weight of a pencil	5 g
Water in a full bathtub	200 L
Water in a washing machine	60 L
Weight of a calculator	100 g
Water in a full kitchen sink	30 L
Weight of a car	1500 kg

Task 8 Problems to solve

1. **(a)** There are now 450 mL in the beaker.
   (b) After the first dilution 500 mL of the 900 mL is the original liquid. This represents $\frac{5}{9}$ of the beaker's contents. The second part of the experiment leaves 450 mL in the beaker. $\frac{5}{9}$ of 450 mL is 250 mL.

2. The last 3 years have been below average rainfall by 70 mm, 110 mm and 275 mm (455 mm in total). So for the 4-year total of rainfall to come back to average, this year we must receive 650 mm plus 455 mm equalling 1105 mm of rainfall.

3. Start with 8.2 kg and subtract the weight of the cardboard carton, leaving 7.9 kg. Now remove 10 × 250 g or 2.5 kg leaving 5.4 kg. Now divide this by 10 to get the result that the beans in each of the cans weigh 540 g.

4. Birth: 3.2 kg. 1 year old: 9.6 kg. 2 years old: 12.8 kg.
   3 years old: 14.4 kg. 4 years old: 16.5 kg. 5 years old: 18.6 kg.

UNIT 12 – IT'S ABOUT TIME
Time

It is interesting to reflect on how many occasions during a day we need to be aware of the time. At my own school the bell rings 16 times in a school day, from 'come in from the yard' time at 8:37am to 'go home' time at 3:30pm. Point out to your students that our lives were not always as ruled by time as they currently are. Prior to the Industrial Revolution, people worked out the time by tracking the sun's passage across the sky and only the wealthy had timing devices.

Ask the students to consider how many terms we use to measure time, from 'second' to 'millennium' and how each term is related to the next in order of size. Discuss the difference between digital and analogue time and the need to be proficient at both forms of time telling. Which type of timing device is now the more prominent? Brainstorming sessions could deal with birthdays, terms and school holidays, time estimation, the use of stopwatches and which time units are fixed due to gravity and physics and which are purely manufactured constructs. You could also discuss the history of timing devices from sundials and 'traditional' clocks to smartphones, watches and so on.

TASK 1

What does it mean?

Understanding the concept of time

Why do you think it is important to know what the time is?

When is it important to know what the time is?

How can we measure what the time is?

TASK 2

The 1-minute challenge

Timing tasks

You will need:

- stopwatch
- BLM 49

Students can work in pairs to complete the following activities:

How often have you heard people say: 'Wait a minute'? This task focuses on what exactly 1 minute really is. For this task you will need a stopwatch. Challenge both yourself and an opponent (or opponents) to complete as many of the following tasks as you can in just 1 minute:

OXFORD UNIVERSITY PRESS

1-minute task	How many I did	Challenger 1's total	Challenger 2's total
Hopping on 1 foot			
Touching your nose, then your tummy			
Stretching high, then touching your toes			
Reading as many words on a page as you can			
Saying the alphabet			
Counting to 10			
Catching a ball			
Clicking fingers			
Saying 'One Cat Dog'			
Counting as far as you can			

This time, try and estimate how long 1 minute actually is. Say 'Start' so your partner knows when to start the stopwatch. When you think 1 minute has gone by say 'Stop' and your partner can record the actual time. Then swap roles. Try and get as close as possible to 1 minute.

My times	How close to 1 minute?	Partner's turns	How close to 1 minute?

Who was closest to 1 minute? How close were they?

TASK 3

Birthday bonanza

Considering the calendar and seasons

You will need:

- calculator

Record the birthdays of 10 people and their current age. They can be people who are close to you and/or you can use your research skills to find the birthdays of some famous people. You may need a calculator to help you with this task.

Person	Birthday	Current age in years
You		
A brother		
A sister		
Mum		
Dad		
A boy cousin		
A girl cousin		
Mum's mum		
Mum's dad		
Dad's mum		
Dad's dad		
Your favourite author		
Your favourite singer		
Your favourite footballer		

TASK 4

Units everywhere

Ranking units of time

You will need:

- BLM 50

1. Complete the table to rank the units of time on the left from shortest to longest. Say something about each unit. The first two have been done for you:

Units	Shortest to longest	Make a statement about each unit
Year	Second	$1 \text{ second} = \frac{1}{60} \text{ minute}$
Season	Minute	$1 \text{ minute} = 60 \text{ seconds}$
Week		$1 \qquad =$

OXFORD UNIVERSITY PRESS

Units	Shortest to longest	Make a statement about each unit
Century		1 =
Minute		1 =
Month		1 =
Millennium		1 =
Second		1 =
Fortnight		1 =
Day		1 =
Semester		1 =
Hour		1 =
Term		1 =
Decade		1 =
Leap year		1 =

2. What is the name given to the length of time it takes for the Earth to spin once on its axis? What is the name given for the length of time it takes for the Earth to orbit the Sun?

3. Before the reign of Julius Caesar there were only 10 months in a year, each longer than the current months. Do research to find out which 2 months the Romans added to the calendar.

TASK 5

Analogue and digital time

Locating and identifying timers

You will need:

- calculator

How many different timing devices can you think of? Where are they found? Are they analogue or digital? List and classify as many timing devices as you can think of.

How many timing devices are there altogether? How many are analogue? How many are digital?

Use a calculator to divide the number of analogue timing devices by the total number of devices. Now multiply this answer by 100, giving you the percentage of devices that are analogue.

Use a calculator to divide the number of digital timers by the total number of timers. Now multiply this answer by 100, giving you the percentage of devices that are digital.

TASK 6

The 10-second reflex challenge

Connecting time and decimals

You will need:

- smartphone

Explain to students that our reflexes enable us to react quickly to a situation, such as catching a ball or avoiding a hot flame.

You can test the quality of your reflexes by using the stopwatch function on a smartphone. Compare the speed of your own reflexes to those of other students in the class. Start a stopwatch and stop it as close to exactly 10 seconds as you can. You will see that the time is given in minutes, seconds and hundredths of a second – the hundredths are shown in decimal format. Record your results and then record the results for other students.

Who had the best single time performance?

TASK 7

The year of the …?

Exploring the Chinese zodiac

You will need:

- BLM 51

Explain to students that people who follow the Chinese calendar believe that the year in which you are born influences your personality and your future. Over a 12-year cycle, a different animal is named as the theme of the year. Those born in that year are believed to have the strengths and weaknesses of that animal. Some animals are seen as being luckier than others. For example, those born in the year of the Rat are believed to be hard working, are good savers of money and love collecting things.

The order of the 12-year cycle is: Rat, Ox, Tiger, Rabbit, Dragon, Snake, Horse, Goat, Monkey, Rooster, Dog and Pig. 2020 was the year of the Rat, 2019 was the year of the Pig, 2018 was the year of the Dog and so on. Complete this table:

Year	Chinese calendar sign	Year	Chinese calendar sign
2020	Rat	2019	Pig
2018	Dog	2017	
2016		2015	
2014		2013	
2012		2011	
2010		2009	
2008		2007	
2001		1994	
1989		1969	
1949		1945	
1914		1901	
1788		1770	

Use your research skills to find famous events that occurred in these years.

TASK 8

TV timing

Exploring duration of time

You will need:

- stopwatch (or smartphone)
- calculator

Pose the following problem to the students:

From free-to-air television, choose four different programs that are of the same length of time (e.g. 30 minutes): three programs from channels with advertisements and one from a channel with no advertisements. Use a stopwatch to find the length of time of each program, and record and compare the results.

Now use a calculator to work out the difference between the advertised time and the actual time. Then multiply this figure by 100 to get the percentage difference.

Now choose your favourite television program from one of these channels and do the same. What is the percentage difference between the advertised time and actual time?

Now watch a movie from one of these channels and do the same. What is the percentage difference between the advertised time and actual time?'

TASK 9
Problems to solve
Applying time skills to unfamiliar situations

1. Are there more or less than a million seconds in a month? Prove your point.

2. Before music streaming and CDs, music was recorded on vinyl discs called 'records'. A single record (with one song on each side) revolved on a turntable at the speed of 45 revolutions per minute. The song 'Imagine' by John Lennon and The Plastic Ono Band runs for 3 minutes and 24 seconds. How many times on a turntable would this single spin for the length of this song?

3. Draw a line splitting this clock face into 2 equal parts in such a way that the numbers on one side of the clock face will sum to the same number on the other side of the clock face:

4. This digital clock shows the time I came home from school last Monday. Note how the hour divides into the minutes without remainder ($36 \div 4 = 9$). How many more times before midnight on Monday will the digital clock show a time when the hour can divide into the minutes without there being a remainder?

ANSWERS

Task 1 What does it mean?

Why? Answers will vary but common responses could be:

   So you are not late for things, so you will be reliable, so we can have a routine, etc.

When? Answers will vary but common responses could be:

   So you can be at an appointment on time, so that you do not miss your favourite television show, so you are not late for school, so a game of sport starts and finishes when it should, so you do not miss a train, etc.

How? Answers will vary but common responses could be:

   Clock, watch, sundial, the sun, water clock, candle clock, stopwatch, shadows, etc.

Task 2 The 1-minute challenge

Answers will vary.

Task 3 Birthday bonanza

Answers will vary.

Task 4 Units everywhere

1.

Shortest to longest	Make a statement about each unit
Second	1 second = $\frac{1}{60}$ minute
Minute	1 minute = 60 seconds
Hour	1 hour = 60 minutes
Day	1 day = 24 hours
Week	1 week = 7 days
Fortnight	1 fortnight = 14 days or 2 weeks
Month	1 month = 28, 29, 30 or 31 days
Term	1 term = about 10 weeks
Season	1 season = 3 months
Semester	1 semester = 2 terms
Year	1 year = 52 weeks, 12 months, 365 days
Leap year	1 leap year = 366 days
Decade	1 decade = 10 years
Century	1 century = 100 years
Millennium	1 millennium = 1000 years

2. The name given to the length of time it takes for the Earth to spin once on its axis is 'day'. The name given for the length of time it takes for the Earth to orbit the Sun is 'year'.

3. The Romans added JULY (named after Julius Caesar) and AUGUST (named after Emperor Augustus).

Task 5 Analogue and digital time

Answers will vary.

Task 6 The 10-second reflex challenge

Answers will vary.

Task 7 The year of the …?

Year	Chinese calendar sign	Year	Chinese calendar sign
2020	Rat	2019	Pig
2018	Dog	2017	Rooster
2016	Monkey	2015	Goat

Year	Chinese calendar sign	Year	Chinese calendar sign
2014	Horse	2013	Snake
2012	Dragon	2011	Rabbit
2010	Tiger	2009	Ox
2008	Rat	2007	Pig
2001	Snake	1994	Dog
1989	Snake	1969	Rooster
1949	Ox	1945	Rooster
1918	Horse	1901	Ox
1788	Monkey	1770	Tiger

Task 8 TV timing

Answers will vary.

Make sure to divide the actual minutes of the program into the advertised length of the program and not vice-versa.

Task 9 Problems to solve

1. There are more than a million seconds in a month.

   $60 \times 60 = 3600$ seconds in an hour. $3600 \times 24 = 86\,400$ seconds in a day.

   $86\,400 \times 12 = 1\,036\,800$ seconds in just 12 days.

2. 1 minute = 45 revolutions per minute (rpm), 3 minutes = 135 rpm.

   There are 45 revolutions every 60 seconds (3 revolutions every 4 seconds). So there must be 18 revolutions in 24 seconds.

   So the single 'Imagine' will run for $135 + 18 = 153$ revolutions.

3. The total of the numbers from 1 to 12 is 78. Half of 78 is 39 so the line must split the clock face into 2 groups of numbers summing to 39: 10, 11, 12, 1, 2, 3 and 4, 5, 6, 7, 8 and 9.

4. There will be 56 more occasions before midnight:

   4 pm: 40, 44, 48, 52, 56.

   5 pm: 05, 10, 15, 20, 25, 30, 35, 40, 45, 50, 55.

   6 pm: 06, 12, 18, 24, 30, 36, 42, 48, 54.

   7 pm: 07, 14, 21, 28, 35, 42, 49, 56.

   8 pm: 08, 16, 24, 32, 40, 48, 56.

   9 pm: 09, 18, 27, 36, 45, 54.

   10 pm: 10, 20, 30, 40, 50.

   11 pm: 11, 22, 33, 44, 55.

OXFORD UNIVERSITY PRESS

UNIT 13 – DOLLARS AND SENSE
Money

Money tends to be one of the topics in mathematics at which students generally excel. I find that the reason for this tends to be twofold. Firstly, the use of money in the community is ubiquitous. Secondly, money tends to be something we are all naturally interested in because we like having it and spending it. With our ever-increasing progress towards a cashless society – electronic banking, credit/debit cards, e-wallets, online purchasing – it will be very interesting to see if this understanding of money continues. Already, some students are losing touch with the common use of money in its physical form.

Remind students that our money system is decimal in nature (based on the number 10), and that money was the first measurement concept to go 'metric' in Australia back in 1966. Ensure that your students are familiar with the coins and notes in our money system and how they relate to each other. Show how amounts of money can be made in many different ways, but that there is only one way of making an amount of money in the most 'economical' manner – that is, using the lowest number of notes and/or coins to make up a given amount of money.

In a brainstorming session, discuss the fact that inflation has resulted, over time, in the removal of smaller value coins (the 1c and 2c coins) and the replacement of smaller notes with coins (the $1 and $2 coins), as money with smaller value becomes less useful. Students can use their research skills as well as their imagination when it comes to money. For example, why were the animals and people on our notes and coins chosen? And what might a $500 note look like in the future?

TASK 1
Know your coins
Recognising Australian coins
You will need:

- Australian coins
- 30 cm ruler

Show students a set of Australian coins.

1. You will see that each coin has our head of state on one side. What or who is on the 'tail' of each coin?

2. How many of these coins would you need to make $2? For example, how many 5c coins make $2?

   5c, 10c, 20c, 50c, $1.

3. Now place a 30 cm ruler on a table. Measure each coin and work out how much money a ruler length (30 cm) of these coins is worth?

   5c, 10c, 20c, 50c, $1, $2

4. What would be the fewest number of coins needed to make the following amounts of money?

   $1.30, 75c, $3.10, 95c, $4.60

TASK 2

Coin creator

Designing a $5 coin

You will need:

- BLM 52

Give students an imaginary scenario: the Australian government decides that it will replace the $5 note with a coin. The students have been invited to enter a national competition to design the 'tail' of this new coin. Give students their imaginary instructions from the government.

Create your coin design in a circle with a 3 cm diameter. You must use native Australian citizens, flora or fauna in the design, but you cannot use any people, plants or animals that have already been used on existing coins. Your coin must have 'five dollars' and the number '5' written on it.

TASK 3

Weighty wallet

Weighing Australian coins

You will need:

- six Australian coins
- scales

Remind the students that the six coins in our money system are different sizes and have different weights. This is useful because when a bank counts the value of a large number of the same coin, it is quicker to weigh the coins than count them.

1. Use scales to find the weight of each of the six Australian coins and round your answer to the nearest gram.

2. Which two coins are closest in weight?

3. Imagine that you have 1 kg (1000 g) of each coin. Approximately how much money would this be worth?

   (Give your answer to the nearest dollar.)

   (Hint: Pretend that there is a $10 coin which weighs 8 g. Work out how many 8s in 1000 – divide 1000 by 8. There are 125, so 125 $10 coins will weigh 1 kg. And 125 $10 coins are worth $1250.)

4. What do you notice about the answers to the 5c, 10c and 20c coins and why do you think this is the case?

TASK 4
Know your notes
Recognising Australian notes
You will need:

- set of Australian notes
- ruler

Show students a set of Australian notes. Ask students to use their research skills to find out some facts about the images shown on the notes:

Did you know that each of our notes has the same width but a different length? Each note is exactly 65 mm in width. Use a ruler to measure their lengths, starting with the $5 note. What do you notice about the length of each note as their value increases?

TASK 5
Note taking
Creating totals using notes
You will need:

- BLM 53

What would be the fewest number of notes needed to make the amounts of money in this table. The first two have been done for you.

Note that the order of the notes does not matter. So, for instance, $10 + $5 is the same answer as $5 + $10.

Amount of money	Different ways of making this amount of money using notes	Number of ways of making this amount of money
$5	$5	1
$10	$10, $5 + $5	2
$15		
$20		

Amount of money	Different ways of making this amount of money using notes	Number of ways of making this amount of money
$25		
$30		
$35		
$40		
$45		
$50		

How many different ways do you think you can make $100 using notes? Check the answers to find out!

TASK 6

A chip off the old block

Exploring unit pricing

You will need:

- calculator

Explain to students that chips (made from potatoes) can be created and purchased in different ways. Challenge students to find the most economical way of buying chips based on the information provided here.

If you can find out the cost of 1 kg of each kind of chips, you will be able to find the cheapest way of getting your chips. (A calculator can be used to check your answers.) Rank your different types of chips from cheapest to most expensive.

1. Buy potatoes and make chips in the kitchen at home:
   Average weight of potato: 200 g
   Cost of potato: 70c
   What is the cost of 1 kg of potatoes?

2. Potato chips in a packet:
   Weight of packet: 125 g
   Cost of packet: $3
   What is the cost of 1 kg of potato chips?

3. Side of chips in a restaurant:
   Weight of chips: 250 g
   Cost of side: $7
   What is the cost of 1 kg of side chips?

4. Takeaway fries at a fast-food restaurant:
   Weight of medium size: 100 g
   Cost of medium fries: $1.80
   What is the cost of 1 kg of takeaway fries?

OXFORD UNIVERSITY PRESS

5. Minimum chips from a fish and chip shop:

Weight of chips: 400 g

Cost of chips: $5

What is the cost of 1 kg of minimum chips?

TASK 7

Currency exchange

Comparing international currencies

You will need:

- BLM 54
- calculator

Remind students that different countries use different currencies, like the yen (Japan) and the peso (e.g. Mexico, Philippines). These currencies are compared with each other regularly and vary in value based on the strength or weakness of worldwide economies.

1. Use your research skills and the website www.xe.com to find the value of 10 Australian dollars (10 AUD) compared to the following international currencies:

Currency	Abbreviation	10 AUD buys
United States dollar		
British pound		
Euro		
New Zealand dollar		
Japanese yen		
Indonesian rupiah		
Thailand baht		
Papua New Guinea kina		
Indian rupee		
Mozambican metical		

2. From this table and your research, which three currencies do you think are stronger than the Australian dollar? Which currency is closest in value to the Australian dollar?

3. Imagine that you go on holiday to the following places and buy breakfast. Using rounding skills and the information you have discovered about currency values, work out the approximate cost of your breakfast in Australian dollars:

Bali breakfast bill: 150 000 IDR

London breakfast bill: 25 GBP

Auckland breakfast bill: 35 NZD

Bangkok breakfast bill: 160 THB

Mumbai breakfast bill: 150 INR

TASK 8

Bean counting

Calculating costs

You will need:

- calculator

Pose the following problem for students:

1. Last week at the supermarket I bought 1 kg of kidney beans, borlotti beans and green beans. My bill came to $9.99. I noticed that the kidney beans were cheaper than the borlotti beans, which, in turn, were cheaper than the green beans. What struck me as being rather amazing was that the cost of the 3 types of beans between them used all the digits 1, 2, 3, 4, 5, 6, 7, 8 and 9.

   There are 3 entirely different ways that this could happen. Can you find all three answers? (Hint: $1.37, $2.96, $8.54 is a good try. It uses all the digits from 1 to 9, but does not add up to *$9.99*.)

2. This week at the supermarket the same purchase of 1 kg of the 3 types of beans cost exactly $9. Again, all the digits from 1 to 9 were used in the bill. What might be the cost of the 3 types of beans this week at the supermarket?

TASK 9

Problems to solve

Applying money skills to unfamiliar situations

1. At Airport Parking the vending machine does not accept credit cards and does not give change. Airport Parking charges in multiples of $5, according to how long you park there. Sarah needs to go to Sydney for 2 days and knows that her parking bill will be at least $100 but not more than $200. So that Sarah will not have to pay extra for her parking, what is the fewest number of notes she will need to have in her purse to be sure that she can make the exact cost of the parking fare?

2. In my wallet I have 9 notes worth $265. I have at least 1 of every note. What is the maximum number of $20 notes that could be in my wallet?

3. In Jack's wallet are notes worth $495. Jack has 1 of 1 type of note, 2 of another type of note, 3 of another type of note, 4 of another type of note and 5 of another type of note. What notes are in Jack's wallet?

4. In the game of Lucky Dip Lotto, contestants dip 10 times into a bag containing an unlimited number of each type of bank note. Thus, the maximum amount of money that can be won is $1000 and the minimum amount is $50. Emma won $300 in Lucky Dip Lotto with 2 of her notes being $100 each. List all the ways that Emma could win $300 with 10 dips which include 2 $100 notes. (Note: The order of the notes does not matter.)

5. 1 Australian dollar is worth about 0.6 Swiss francs. How much Australian money would someone from Switzerland receive if they exchanged 100 Swiss francs?

OXFORD UNIVERSITY PRESS

ANSWERS

Task 1 Know your coins

1. From left to right: an echidna, a lyrebird, a platypus, an emu and kangaroo on either side of the Australian coat of arms, 5 kangaroos, an Aboriginal and Torres Strait islander elder (based on Gwoya Tjungurrayi).

2. 5c: 40, 10c: 20, 20c: 10, 50c: 4, $1: 2.

3. Answers may vary slightly, depending on the length of the '30 cm' ruler. Rulers usually both start before 0 cm and end after 30 cm.

   5c: 75c or 80c, 10c: $1.20 or $1.30, 20c: $2 or $2.20, 50c: $4.50 or $5, $1: $12 or $13, $2: $28 or $30.

4. $1.30 = 3, 75c = 3, $3.10 = 3, 95c = 4, $4.60 = 4

Task 2 Coin creator

Answers will vary.

Task 3 Weighty wallet

1. Rounded to the nearest gram:

   5c: 3 g (2.83 g), 10c: 6 g (5.65 g), 20c: 11 g (11.3 g), 50c: 16 g (15.55 g), $1: 9 g (9 g), $2: 7 g (6.6 g).

2. The 2 coins closest in weight are the 10c and $2 coins.

3. 1 kg of each coin would be worth:

   5c: $17 ($18), 10c: $17 ($18), 20c: $18 ($18), 50c: $31 ($32), $1: $111 ($111), $2: $286 ($303).

   (Note: The first answer uses the rounded weight of each coin to the nearest gram. The total in brackets is the exact answer.)

4. 1 kg of 5c, 10c and 20c coins are all worth the same because as their value doubles so does their weight.

Task 4 Know your notes

$5: Elizabeth II, the queen; Parliament House, Canberra.

$10: Banjo Paterson, poet; Dame Mary Gilmore, writer and journalist.

$20: Reverend John Flynn, started the Inland Mission and the Royal Flying Doctor Service; Mary Reibey, businesswoman and merchant.

$50: David Unaipon, preacher, author and inventor; Edith Cowan, first woman in an Australian parliament.

$100: Sir John Monash, First World War army general; Dame Nellie Melba, opera singer.

Note lengths: $5: 130 mm, $10: 137 mm, $20: 144 mm, $50: 151 mm, $100: 158 mm. The notes get 7 mm longer each time.

Task 5 Note taking

$15 = $10 + $5, $5 + $5 + $5 (2 ways).

$20 = $20, $10 + $10, $10 + $5 + $5, $5 + $5 + $5 + $5 (4 ways).

$25 = $20 + $5, $10 + $10 + $5, $10 + $5 + $5 + $5, $5 + $5 + $5 + $5 + $5 (4 ways).

$30 = $20 + $10, $20 + $5 + $5, $10 + $10 + $10, $10 + $10 + $5 + $5, $10 + $5 + $5 + $5 + $5, $5 + $5 + $5 + $5 + $5 + $5 (6 ways).

$35 = $20 + $10 + $5, $20 + $5 + $5 + $5, $10 + $10 + $10 + $5, $10 + $10 + $5 + $5 + $5, $10 + $5 + $5 + $5 + $5 + $5, $5 + $5 + $5 + $5 + $5 + $5 + $5 (6 ways).

$40 = $20 + $20, $20 + $10 + $10, $20 + $10 + $5 + $5, $20 + $5 + $5 + $5 + $5, $10 + $10 + $10 + $10, $10 + $10 + $10 + $5 + $5, $10 + $10 + $5 + $5 + $5 + $5, $10 + $5 + $5 + $5 + $5 + $5 + $5, $5 + $5 + $5 + $5 + $5 + $5 + $5 + $5 (9 ways).

$45 = $20 + $20 + $5, $20 + $10 + $10 + $5, $20 + $10 + $5 + $5 + $5, $20 + $5 + $5 + $5 + $5 + $5, $10 + $10 + $10 + $10 + $5, $10 + $10 + $10 + $5 + $5 + $5, $10 + $10 + $5 + $5 + $5 + $5 + $5, $10 + $5 + $5 + $5 + $5 + $5 + $5 + $5, $5 + $5 + $5 + $5 + $5 + $5 + $5 + $5 + $5 (9 ways).

$50 = $50, $20 + $20 + $10, $20 + $20 + $5 + $5, $20 + $10 + $10 + $10, $20 + $10 + $10 + $5 + $5, $20 + $10 + $5 + $5 + $5 + $5, $20 + $5 + $5 + $5 + $5 + $5 + $5, $10 + $10 + $10 + $10 + $10, $10 + $10 + $10 + $10 + $5 + $5, $10 + $10 + $10 + $5 + $5 + $5 + $5, $10 + $10 + $5 + $5 + $5 + $5 + $5 + $5, $10 + $5 + $5 + $5 + $5 + $5 + $5 + $5 + $5, $5 + $5 + $5 + $5 + $5 + $5 + $5 + $5 + $5 + $5 (13 ways).

$100 can be made in 50 different ways.

Task 6 A chip off the old block

Different types of chips ranked cheapest to most expensive:

1. Chips made at home: $3.50 per kg
5. Minimum chips from fish and chip shop: $12.50 per kg
4. Takeaway fries: $18 per kg
2. Potato chips in a packet: $24 per kg
3. Side of chips in restaurant: $28 per kg.

Task 7 Currency exchange

1. United States dollar: USD, approximately 6.35 USD
   British pound: GBP, approximately 5.08 GBP
   Euro: EUR, approximately 5.85 EUR
   New Zealand dollar: NZD, approximately 10.55 NZD
   Japanese yen: JPY, approximately 684 JPY
   Indonesian rupiah: IDR, approximately 98 000 IDR
   Thailand baht: THB, approximately 206 THB
   Papua New Guinea kina: PGK, approximately 22 PGK
   Indian rupee: INR, approximately 485 INR
   Mozambican metical: MZM, approximately 427 MZN.

2. The 3 currencies stronger than Australia's are the US dollar, Euro and British pound. The currency closest in value to Australia's is the New Zealand dollar.

3. Breakfasts would cost approximately: Bali $15, London $50, Auckland $35, Bangkok $8, Mumbai $3.

Task 8 Bean counting

1. Kidney beans $1.52, borlotti beans $3.68, green beans $4.79.

Kidney beans $1.84, borlotti beans $2.36, green beans $5.79.

Kidney beans $1.69, borlotti beans $2.83, green beans $5.47.

2. Kidney beans $1.78, borlotti beans $2.53, green beans $4.69.

(There are variations of these answers which transfer the debts to other bean types.)

Task 9 Problems to solve

1. 6 notes: $100, $50, $20, $20, $10, $5.

2. 5 × $20 notes: the 9 notes could be $100, $50, $20, $20, $20, $20, $20, $10, $5.

3. 2 × $100, 4 × $50, 3 × $20, 1 × $10, 5 × $5.

4. 5 ways:

$100, $100, $50, $20, $5, $5, $5, $5, $5, $5.

$100, $100, $50, $10, $10, $10, $5, $5, $5, $5.

$100, $100, $20, $20, $20, $20, $5, $5, $5, $5.

$100, $100, $20, $20, $20, $10, $10, $10, $5, $5.

$100, $100, $20, $20, $10, $10, $10, $10, $10, $10.

5. If 1 Australian dollar buys 0.6 Swiss francs then 1 Swiss franc must buy the reciprocal of $\frac{6}{10}$ or $\frac{10}{6}$ Australian dollars. $\frac{\$10}{6}$ equals $\frac{\$5}{3}$ or $\frac{\$1-2}{3}$ or $1.66. Thus 100 Swiss francs must buy $166.66 Australian dollars.

UNIT 14 – DATA DRIVEN
Statistics and graphs

Gathering and analysing data and interpreting graphs and tables have never been so important as they are today. News sources and advertisements for products are full of claims and results that are supposedly 'data driven'. It is vital that our students can see beyond such 'noise' and critically analyse what is regularly presented to them in statistical form. The COVID-19 pandemic is a wonderful example of the vital nature of data analysis, with trends and patterns being drawn constantly from numbers and graphs, across local government areas, states and nations.

Students love gathering data and conducting surveys from a very young age. It is the analysis of this data that leads to creative and critical thinking. Drawing conclusions, interpreting results and looking for patterns leads to the foundational skills that are so important for problem solving. When brainstorming for this unit, use various sources to show where and how data is displayed. Discuss sporting contexts where data is becoming more and more useful when analysing performances. What does Sir Donald Bradman's test batting average of 99.94 mean? How do we know that October is the wettest month in Melbourne? How can a line graph show a trend?

Remind students that data can be represented in many different forms (pictographs, bar and column graphs, line graphs, pie graphs, two-way tables and Venn diagrams) and we can use statistical analysis of the mean, mode, median and range of data sets to draw meaningful conclusions.

TASK 1
What does it mean?
Understanding data and graphs

We gather data regularly to look for patterns or trends to help us draw conclusions and plan for the future. Graphs are basically pictures, or visual representations, drawn from the data to help us understand what the data tells us.

Discuss examples of when and why we use data, and the different ways it can be presented:

1. Can you think of examples of where and when data has been gathered to show patterns or trends?

2. Can you think of examples of where graphs have been used to show patterns or trends?

3. Can you think of different types of graphs that can be used to show patterns or trends?

OXFORD UNIVERSITY PRESS

TASK 2

Draw it

Making a pictograph

You will need:

- BLM 55

Remind students that pictographs use pictures or icons to represent data. They are the simplest type of graphs and are the easiest to interpret.

Types of fruit

1. This pictograph shows how much fruit 11 students brought to school for lunch.

   How many bananas were brought to school for lunch?

   How many apples were brought to school for lunch?

   How many cherries were brought to school for lunch?

   How many pieces of fruit were brought altogether?

2. Work with five friends to create a pictograph. Find out the number of televisions each of you have in your home. Record the data and then use the data to create a pictograph.

Name	Number of televisions
Me	

TASK 3

Stepping out

Making a bar graph

You will need:

- BLM 56

Explain to students that bar graphs use rows to set out and compare data horizontally:

1. This bar graph shows the results of a music survey:

    (a) What question do you think was asked in the survey?

    (b) Which type of music was least liked?

    (c) How many people answered the survey?

2. Work with four friends to create a bar graph, showing how many steps you take to walk around a sports oval or a basketball court. Record your names on the left. At the bottom of the graph, you can see the number of steps taken from 0 to 1000. Round your answers to the nearest 100 steps and colour in the rows to show the data.

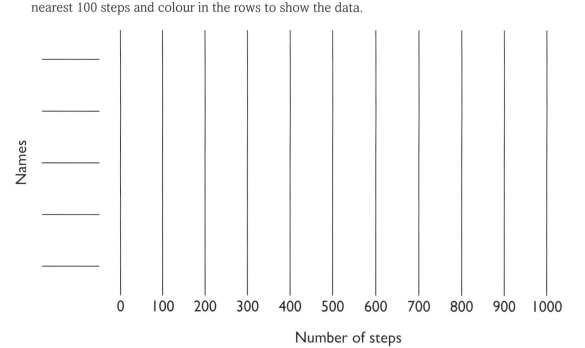

TASK 4

Roll the dice

Making a column graph

You will need:

- BLM 57
- dice

Explain to students that column graphs are bar graphs with a twist – they use columns to set out data vertically.

OXFORD UNIVERSITY PRESS

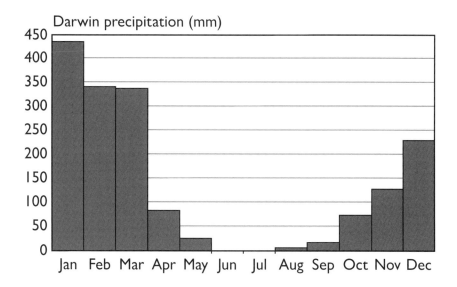

Darwin precipitation (mm)

1. This column graph shows the results of a survey:
   (a) What do you think this graph is showing?
   (b) What is the wettest season in Darwin?
   (c) Melbourne receives about 650 mm of rain per year. In Darwin, in which two consecutive months does the city receive more rainfall than Melbourne does in a year? (There are three different answers.)

2. **(a)** Make a column graph. Roll a 6-sided die twice and then add up the two numbers that you roll (e.g. 4 then 2 will make 6). Colour in a square each time you roll the numbers listed along the bottom. Keep rolling and recording until at least one number is rolled ten times.

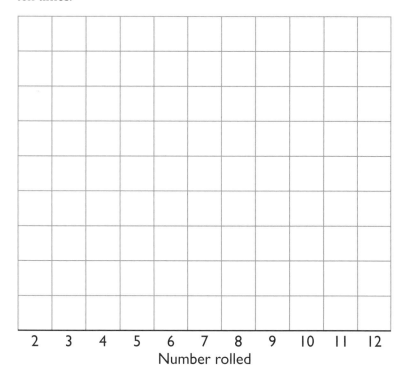

Number rolled

   (b) What patterns or trends did you notice occurring in this task?
   (c) Why do you think this happened?

TASK 5

Minimum temperatures

Making a line graph

You will need:

- BLM 58

Explain to students that line graphs are designed to show changes over time. They use dots to record data and a line to join the dots, showing changes in the data over time.

1. What do you think this graph is showing?

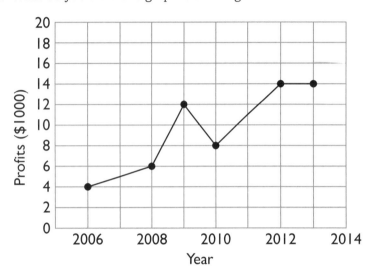

2. In which year did profits fall for the company?

3. What was the most profitable year for the company?

4. What happened to profits between 2012 and 2013?

5. Make a line graph. Record the minimum temperature where you live for one week, from a Sunday to the following Saturday.

TASK 6

A slice of pie

Making a pie graph

You will need:

- BLM 59

Remind students that pie graphs are also known as pie charts or sector graphs. They display data in sections, like the slices of a pie or a pizza. Pie graphs often use percentages to explain the data they show. The total of the 'pie' is 100%.

1. A large number of people were surveyed about their favourite colour and the results were shown on a pie graph:

   (a) List the colour choices in order from least to most popular.

   (b) Which two colours that are next to each other come closest to 50% of the votes?

   (c) Which colour comes closest to 25% of the votes?

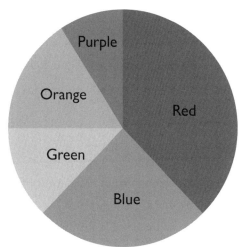

2. Ask 9 friends or family members to name their favourite Australian state or territory for a holiday.

   Their choices are: Australian Capital Territory (ACT), New South Wales (NSW), Northern Territory (NT), Queensland (QLD), South Australia (SA), Tasmania (TAS), Victoria (VIC) or Western Australia (WA).

   This will give you 9 votes and each vote will represent 40 degrees of the circle. Use a protractor and a ruler to complete the pie chart. Each state or territory in your pie chart will need to be coloured in with a different coloured pencil. If any state or territory does not get a vote, do not include it on the pie chart.

Survey vote results:

ACT: NSW: NT: QLD: SA: TAS: VIC: WA:

Favourite state or territory

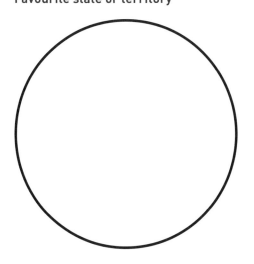

TASK 7

Two-way tables

Working with more than one variable

You will need:

- BLM 60

Explain to students that two-way tables display data that contains two or more variables or options. These tables can show data that overlaps.

1. This two-way table shows sports played by some Year 5 students:

Sports played	Boys	Girls
Soccer	34	37
Tennis	18	11
Do not play either	16	24

(a) How many boys and how many girls were in the survey?

(b) How many students in Year 5 play tennis?

(c) How many students in Year 5 do not play soccer?

2. At Keynote Primary School, all students in Years 5 and 6 must play a string or woodwind instrument. This two-way table shows the number of students in Year 5 and Year 6 who play a string or woodwind instrument:

	Year 5 students	Year 6 students
String instrument	25	16
Woodwind instrument	22	34
Play both	11	17

(a) How many students in Year 5 play a woodwind instrument?

(b) How many students play a string instrument in Years 5 and 6 altogether?

(c) How many students play more than one instrument in Year 6?

(d) How many more students are in Year 6 than Year 5 at Keynote Primary School?

3. Survey the students in your class. How many girls travel to school by private transport (e.g. car, bicycle) and how many travel by public transport (e.g. bus, train). How many boys? Show the results in a two-way table:

Type of transport	Girls	Boys
Private transport		
Public transport		
Both types		

TASK 8
Overlapping data
Making a Venn diagram

Explain to students that Venn diagrams are a way of showing how two or more sets of data are connected. They were created by the English mathematician John Venn in the 19th century. Venn diagrams use two or more circles, which often overlap, to show how groups of results are related.

1. This Venn diagram shows the results of a survey conducted with 100 students. It shows whether or not they ate an apple (A) or a banana (B) yesterday. There are four groups of data, each one expressed as a decimal.

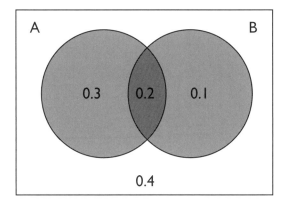

   (a) What do you think the 0.4 section stands for?
   (b) What do you think the 0.3 section stands for?
   (c) What do you think the 0.2 section stands for?
   (d) What do you think the 0.1 section stands for?
   (e) How many students ate an apple yesterday?
   (f) How many students ate a banana yesterday?
   (g) How many students ate both an apple and a banana yesterday?
   (h) How many students ate neither an apple nor a banana yesterday?

2. This Venn diagram shows the relationship between three different sets of data. In Year 6, students must choose at least one sport to play. They can choose two sports if they wish, even three if they are very keen. The sports offered are athletics (A), basketball (B) and cricket (C).

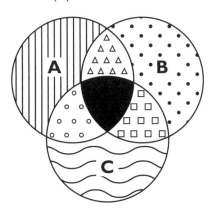

(a) Which sport or sports were chosen by students in the striped section?

(b) Which sport or sports were chosen by students in the triangle section?

(c) Which sport or sports were chosen by students in the dotty section?

(d) Which sport or sports were chosen by students in the circle section?

(e) Which sport or sports were chosen by students in the black section?

(f) Which sport or sports were chosen by students in the square?

(g) Which sport or sports were chosen by students in the wavey-lined section?

3. Can you think of a survey question that you could ask your classmates that would result in data sets that would have an overlap, like we see in the Venn diagrams found above?

TASK 9

Better than average

Interpreting sporting data

You will need:

- BLM 61

Explain to students that when we find the average of a set of data, we get an overall picture of what that data is telling us. For example, averages are used regularly to show how wet or dry a month, season or year has been, how well a company is doing this year compared to last year and how well a class does on a test.

1. Sport is full of statistics and averages. In cricket, batters are often ranked by their average batting score. The total amount of runs they have made is divided by the number of times they have been out. Calculate the batting averages for 10 performances by these batters. Use a calculator and round your answers to two decimal places.

Batter's name	Last 10 scores in tests	Times out	Average
Babar Azam (Pakistan)	5, 47, 11, 63, 7, 30, 77, 8, 0, 2	9	
Virat Kohli (India)	3, 14, 74, 4, 11, 72, 0, 62, 27, 0	10	
Mithali Raj (India)	47, 0, 65, 22, 4, 50, 37, 2, 4, 30	9	
Marnus Labuschagne (Australia)	215, 59, 47, 6, 48, 28, 91, 73, 108, 25	10	
Cheteshwar Pujara (India)	50, 77, 25, 56, 73, 15, 21, 7, 0, 17	10	
Ajinkya Rahane (India)	22, 4, 37, 24, 1, 0, 67, 10, 7, 27	10	
Joe Root (England)	186, 11, 218, 40, 6, 33, 17, 19, 5, 30	10	
Jane Smit (England)	56, 33, 19, 2, 4, 21, 12, 0, 15, 23	7	
Steve Smith (Australia)	7, 63, 1, 1, 0, 8, 131, 81, 36, 55	9	
Ben Stokes (England)	0, 9, 82, 7, 18, 8, 6, 25, 55, 2	10	
David Warner (Australia)	43, 19, 41, 38, 45, 111, 5, 13, 1, 48	9	

OXFORD UNIVERSITY PRESS

Batter's name	Last 10 scores in tests	Times out	Average
Kane Williamson (New Zealand)	14, 9, 0, 89, 3, 5, 251, 129, 21, 238	10	

2. Now rank these batters by their average in order from 1st to 10th.

TASK 10

Mean, mode, median and range

Using statistics

You will need:

- dice

Explain to students that there are four common statistical tests that can be applied to data, to produce useful information. The first is the mean, and in this task, the mean will be considered the same as the average (although this is not always the case). The mode of a set of data is the item that is found the most, and there is often more than one mode in a data set. The median in a set of data is the item that is found in the middle when the items are ranged from smallest to largest. If there are an even number of items in the data set, the average of the two items found in the middle is the median. The range of a set of data is the difference between the smallest and the biggest values.

1. Michael Jordan is widely regarded as being the greatest basketballer ever. This table shows the 10 highest scores he made in his NBA career playing for the Chicago Bulls. Work out the following for the 10 scores:

(a) mean (average)

(b) mode(s)

(c) median

(d) range.

Points	Season	Opponents
69	1989–1990	Cavaliers
64	1992–1993	Magic
63	1985–1986	Celtics
61	1986–1987	Hawks
61	1986–1987	Pistons
59	1987–1988	Pistons
58	1986–1987	Nets
57	1992–1993	Bullets
56	1991–1992	Heat
56	1986–1987	76ers

2. Roll a 6-sided die 30 times and record the numbers you roll.

   Work out the: (a) mean (average), (b) mode(s), (c) median and (d) range of the data you have gathered.

3. The mean (average) of the numbers that can be expected to be rolled on a 6-sided die is 3.5. Because every number on the dice has an equal chance of being rolled, this can be calculated by adding up all the possibilities (1, 2, 3, 4, 5, 6) and dividing by 6. Now look at the mean you calculated for your data set. How close was it to 3.5?

TASK 11
Problems to solve
Applying data skills to unfamiliar situations

1. In the first 5 games of the season the Brolgas netball club averaged 31 goals per game. After game 6 of the season their average number of goals per game went up to 33. How many goals did the Brolgas score in game 6?

2. The median of 7 numbers is 14, and the average of the 7 numbers is 15. What could be the biggest possible number in this data set?

3. The mode of 6 numbers is 3. The biggest of the 6 numbers is 22. What is the biggest possible average that the numbers could have?

4. On a maths test, the range of the results was 45 marks. No student scored 100%. 5 students were in the 90s. 10 students sat the test. All students scored different whole number scores. What is the biggest possible average score for the 10 results?

5. 5 girls sat a 10-question grammar test. The average score for the 5 results was 7.9. The mode was 10.

   The median was 7. The range was 4.

   What were the 5 results?

ANSWERS

Task 1 What does it mean?

Answers will vary but may include:

1. and 2. When playing games of chance, when analysing sporting performances, when looking at companies and their profits, when looking at the results of a census, when looking at rainfall or temperatures, when looking at pandemic rates of infection, when looking at unemployment data, when looking at population statistics, etc.

3. Tables, pictographs, bar graphs, column graphs, line graphs, pie charts.

Task 2 Draw it

1. 4 bananas, 3 apples, 8 cherries, 15 pieces of fruit altogether.

2. Answers will vary.

Task 3 Stepping out

1. (a) Do you like hip hop, classical, rock or jazz music the most?

(b) Rock music was the least favourite.

(c) There were 41 people who answered the survey.

2. Graph results will vary.

Task 4 Roll the dice

1. (a) Average monthly annual rainfall for Darwin.

(b) The wettest season in Darwin is summer.

(c) In December and January, in January and February and in February and March, Darwin receives more rainfall than Melbourne does in a whole year.

2. (a) Graph results will vary.

(b) The column graph will look like a triangle with the apex at 6, 7 or 8.

(c) This is because of the 36 ways of rolling a die twice: 1 adds up to 2, 2 add up to 3, 3 add up to 4, 4 add up to 5, 5 add up to 6, 6 add up to 7, 5 add up to 8, 4 add up to 9, 3 add up to 10, 2 add up to 11 and 1 adds up to 12. So 16 rolls in every 36 should sum to 6, 7 or 8.

Task 5 Minimum temperatures

1. A company's profits from 2006 to 2013.

2. Profits fell between 2009 and 2010.

3. Profits were highest between 2012 and 2013.

4. Profits were flat.

5. Graph results will vary.

Task 6 A slice of pie

1. (a) Purple, green, orange, blue, red.

(b) Red and purple.

(c) Blue.

2. Pie graph results will vary.

Task 7 Two-way tables

1. (a) 68 boys and 72 girls were in the survey.

(b) 29 students in Year 5 play tennis.

(c) 69 students in Year 5 do not play soccer.

2. (a) 33 students in Year 5 play a woodwind instrument.

(b) 69 students in Years 5 and 6 play a string instrument.

(c) 17 students in Year 6 play more than 1 instrument.

(d) There are 9 more students in Year 6 at Keynote Grammar.

3. Survey results will vary.

Task 8 Overlapping data

1. **(a)** The students who did not eat either an apple or a banana yesterday.
   (b) The students who ate an apple but not a banana yesterday.
   (c) The students who ate both an apple and a banana yesterday.
   (d) The students who ate a banana but not an apple yesterday.
   (e) 50 students.
   (f) 30 students.
   (g) 20 students.
   (h) 40 students.

2. **(a)** Athletics.
   (b) Athletics and basketball.
   (c) Basketball.
   (d) Athletics and cricket.
   (e) Athletics, basketball and cricket.
   (f) Basketball and cricket.
   (g) Cricket.

3. Answers will vary.

Task 9 Better than average

1.

Batter's name	Last 10 scores in tests	Times out	Average
Babar Azam (Pakistan)	5, 47, 11, 63, 7, 30, 77, 8, 0, 2	9	27.78
Virat Kohli (India)	3, 14, 74, 4, 11, 72, 0, 62, 27, 0	10	26.70
Mithali Raj (India)	47, 0, 65, 22, 4, 50, 37, 2, 4, 30	9	29.00
Marnus Labuschagne (Australia)	215, 59, 47, 6, 48, 28, 91, 73, 108, 25	10	70.00
Cheteshwar Pujara (India)	50, 77, 25, 56, 73, 15, 21, 7, 0, 17	10	34.10
Ajinkya Rahane (India)	22, 4, 37, 24, 1, 0, 67, 10, 7, 27	10	19.90
Joe Root (England)	186, 11, 218, 40, 6, 33, 17, 19, 5, 30	10	56.50
Jane Smit (England)	56, 33, 19, 2, 4, 21, 12, 0, 15, 23	7	26.43
Steve Smith (Australia)	7, 63, 1, 1, 0, 8, 131, 81, 36, 55	9	42.56
Ben Stokes (England)	0, 9, 82, 7, 18, 8, 6, 25, 55, 2	10	21.20
David Warner (Australia)	43, 19, 41, 38, 45, 111, 5, 13, 1, 48	9	40.44
Kane Williamson (New Zealand)	14, 9, 0, 89, 3, 5, 251, 129, 21, 238	10	75.90

2. 1st. Williamson, 2nd: Labuschagne, 3rd: Root, 4th: Smith, 5th: Warner, 6th. Pujara, 7th: Azam, 8th: Kohli, 9th: Stokes, 10th: Rahane.

Task 10 Mean, mode, median and range

1. **(a)** Mean = 60.4 points.

   (b) Modes = 61 and 56 points.

   (c) Median = 60 points.

   (d) Range = 13 points.

2. Answers will vary.

3. Answers will vary.

Task 11 Problems to solve

1. If the Brolgas averaged 31 goals per game after 5 games they must have scored 5 × 31 (155) goals for the season. An average of 33 goals after game 6 increases this total to 6 × 33 (198) goals.

   198 − 155 = 43 goals scored in game 6.

2. The total of the 7 numbers must be 7 × 15 = 105. So 14 must be the 4th number when listed from smallest to biggest. Thus the first 6 numbers could be 1, 1, 1, 14, 14, 14, with the biggest possible being 60.

3. The numbers could be 3, 3, 19, 20, 21, 22, giving an average of 14.67.

4. The scores could have been 99, 98, 97, 96, 95, 89, 88, 87, 86 and (99 − 45) 54. This gives a total of 889 and an average of 88.9 marks.

5. From the information given we know 4 of the 5 results: 10, 10, 7, 6. If the average is 7.8, we know the total of the 5 scores must be 7.9 × 5 = 39.5. The 4 known results sum to 33, thus the 5th result must be 6.5.

UNIT 15 – SHAPING UP
Geometry

The word 'geometry' comes from two Greek words: 'geo' meaning 'the Earth' and 'metro' meaning 'to measure'. It was during the Golden Age of Greece (500–300 BCE) that mathematicians first started to explore the geometric nature of the world by studying two-dimensional (2D) shapes, three-dimensional (3D) objects, directions and angles.

Students love exploring geometric concepts and the myriad of excellent concrete materials available to them helps lead to a deep foundational understanding that is so vital for the formal work in the topic that will come in secondary school. Pattern blocks tessellate beautifully and demonstrate that hexagons can be formed by fitting together two identical trapeziums, three rhombuses and six equilateral triangles; geoshapes connect to create a vast range of polyhedrons; and clinometers help build the concept and meaning of 'degree' and 'angle'.

Like the ancient Greeks, encourage your primary students to explore the geometric world around them by hunting for tessellated polygons in the environment, 2D shapes, 3D objects and angles in the school and at home. Brainstorming sessions should introduce polygon relationships, 3D and 2D connections and commonly found angles. Simple computer codes often include angles and directions too, leading to the importance of precise instructional language and unambiguous vocabulary.

TASK 1
What does it mean?
Understanding the concept of geometry

The word 'geometry' comes from two ancient Greek words: 'geo' meaning the Earth and 'metro' meaning 'to measure'. So it was through the study of directions, angles, two-dimensional (2D) shapes and three-dimensional (3D) objects that the ancient Greeks were able to make sense of the structure of the world around them.

1. Where and when might we need to use directions?
2. Where and when might we see or use angles?
3. Where and when might we see or use 2D shapes?
4. Where and when might we see or use 3D objects?

TASK 2
Robots
Using directions

Explain to students that in this task, they will work with a partner. One student will be a scientist and the other will be the robot that has been made by the scientist in a laboratory. The scientist's job is to program the robot to move around a room or house as efficiently as possible. The robot must follow the instructions of the scientist and do exactly as programmed.

Start by moving around a small area and then, when you get better, stretch out the area in which you are working. Take it in turns to be the robot and the scientist. Some examples of programmed instructions that the scientist might say are:

> *'Move forward 10 steps.'*
> *'Turn left.'*
> *'Go back 2 steps.'*
> *'Make a quarter-turn right.'*
> *'Go up 12 stairs.'*

TASK 3

Architect and builder

Using positional language

You will need:

- building bricks

Explain to students that in this task, they will work with a partner. One student will be an architect and the other will be a master builder. The architect and builder will have the same number of different-coloured building bricks (e.g. 1 green, 1 blue, 1 yellow, 1 red, 1 white and 1 black brick). The students sit back to back so that they cannot see what the other is doing. The architect then builds anything that they like with the bricks that they have. The architect must describe to the master builder what has been made and give step-by-step instructions of how to build it. The master builder then shows the architect their creation to check how accurately it has been made.

Remind students that the architect can use words such as 'connect', 'join', 'left', 'right', 'above', 'below', 'turn', 'it looks like'.

As students get better at this game, increase the number of bricks in the creation to make the job of each player harder.

TASK 4

Compass points

Using directions

You will need:

- compass
- BLM 62

Explain to students how a compass works. Inside a compass is a small magnet. The needle on a compass reacts to the magnetic field of the Earth and always points north. Smartphones and many fitness watches contain a compass. Students can use a compass to find the answers to these questions:

1. Answer the following:
   (a) What direction does the front of your house or classroom face?
   (b) What direction does your bedroom window or classroom window face?

(c) What direction does your bedroom door or classroom door face?

(d) Go for a walk outside, record the directions you travelled.

2. Now pretend that you are a fly sitting in the middle of the ceiling of your bedroom or classroom.

   In the rectangle draw a sketch of what the fly would be looking at and the direction from the fly that your bed, the window, the door, the bedside tables and the wardrobe would be.

TASK 5

Angles all around

Identifying angles

You will need:

- protractor

Remind students that an angle is formed whenever two lines meet. A common angle in building and construction is the right angle. It measures 90 degrees and looks like a corner where two lines meet or the letter L (forwards, backwards and upside down).

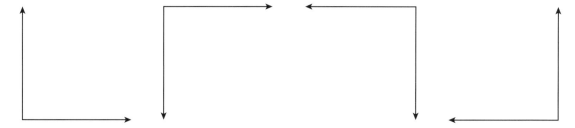

1. Find and record 10 right angles in your local environment (e.g. the classroom, the schoolyard or at home).

   The most common angle in the natural world is the acute angle. This looks like the letter 'V' (forwards, backwards, sideways) and measures less than 90 degrees.

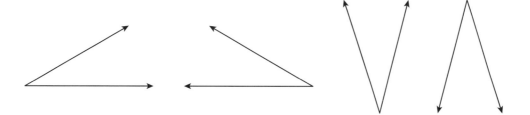

2. Steep slopes are examples of acute angles – they are called inclines when they go up and declines when they go down. The world's steepest street, Baldwin Street, is found in the city of Dunedin, in New Zealand's South Island. Can you find out how many degrees it measures?

3. Find 10 acute angles in your local environment (e.g. in the classroom, the schoolyard or at home).

4. Here are the 26 letters of the alphabet shown as capitals:

   A B C D E F G H I J K L M N O P Q R S T U V W X Y Z

   (a) Which of these letters contain a right angle?

   (b) Which of these letters contain an acute angle?

   (c) Which of these letters contain neither a right angle nor an acute angle?

TASK 6

Mirror image

Working with reflections and symmetry

You will need:

- mirror
- BLM 63

Explain to students that a shape can be transformed in different ways: it can be rotated (turned), dilated or compressed (enlarged or reduced) or reflected (flipped).

When an image is reflected in a mirror, it is as though a line of symmetry is used to split the image so that what is on the left can be folded to match what is on the right.

1. In the table, can you tell which letters or numbers would appear the same when they are reflected? Use a mirror to check your answers:

Image	Flip	Image	Flip	Image	Flip	Image	Flip	Image	Flip	Image	Flip
A		B		C		D		E		F	
G		H		I		J		K		L	
M		N		O		P		Q		R	
S		T		U		V		W		X	
Y		Z		1		2		3		4	
5		6		7		8		9		10	

2. MUM is a word that appears the same when reflected in a mirror. Are the following words the same when reflected? Answer 'yes' or 'no':

   (a) DAD

   (b) LEVEL

   (c) TOOT

   (d) BOB

   (e) HANNAH

3. Draw a picture in the left half of the grid and then reflect it on the right side of the grid:

8	7	6	5	4	3	2	1	1	2	3	4	5	6	7	8

TASK 7

Co-ordinates

Plotting points on a grid

You will need:

- BLM 64

Explain to students that about 400 years ago, a French mathematician named René Descartes created what is now called 'co-ordinate geometry' or, in his honour, Cartesian geometry. It was a way of showing position on a grid and eventually led to the creation of map co-ordinates and even Global Positioning System (GPS) technology.

The grid that Descartes created began at the bottom left corner in a place called the 'origin'. From the origin, evenly spaced horizontal and vertical lines create evenly spaced squares. Where the lines meet, points are drawn or 'plotted' and can be joined to create shapes.

From the origin, the horizontal lines (x co-ordinates) were counted before the vertical lines (y co-ordinates). This made what is now known as an 'ordered pair'. An example of an ordered pair is (5,7) and refers to a position where 5 lines horizontally and 7 lines vertically from the bottom left corner meet. Note the brackets around the ordered pair and the comma separating the co-ordinates.

1. On this grid, 3 points have been plotted and joined to form a triangle. Can you list the 3 co-ordinate points that represent the 3 vertices (corners) of the triangle?

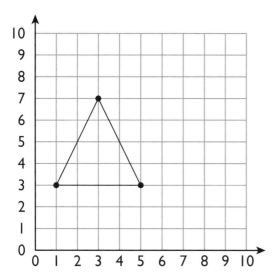

2. On the grid in BLM 64 create your own shape. You must use 6 different points and join them up to create a hexagon. Remember that not all hexagons are regular and have equal lengths. On the grid, write A, B, C, D, E and F near the 6 points. Then write below the grid the co-ordinate points which are at the vertices of your hexagon:

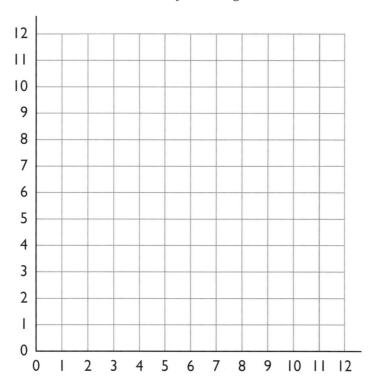

Co-ordinate points used: A =   B =   C =   D =   E =   F =

TASK 8

2D shapes

Recognising polygons and their angles

You will need:

- BLM 65

Explain to students that the word 'polygon' comes from two ancient Greek words: 'poly' meaning 'many' and 'gon' meaning 'corners'. Thus, a polygon is literally a shape with many corners. Polygons have a perimeter and can be drawn on a flat surface or a plane. The simplest polygon is a triangle.

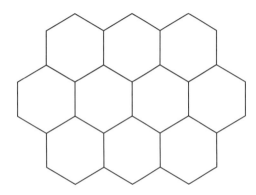

1. When polygons are fitted together without gaps they are tessellated. Inside and outside your house or school you can see many examples of tessellated shapes. They may be found in tiles or possibly paintings. List places in your house or school where you can find tessellated polygons and the type of polygon that has been tessellated.

2. Use your research skills to find the names of polygons with a certain number of vertices. Some you may already know.

Vertices	Polygon name	Vertices	Polygon name	Vertices	Polygon name
3		4		5	
6		7		8	
9		10		11	
12		13		14	
15		16		17	
18		19		20	

3. Any triangle contains 180 degrees. The number of degrees found in a polygon is determined by the number of triangles that can be found within it. Each triangle in this diagram is made up of a right angle and two 45 degree angles that have been formed by splitting two right angles. You can work out how many degrees in a polygon by using one vertex of that polygon and drawing lines to its other vertices to form as many triangles as you can.

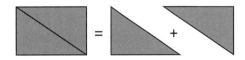

OXFORD UNIVERSITY PRESS

This next diagram shows how you can do this to see that there are four possible triangles in a hexagon.

4 × 180 degrees = 720 degrees, the number of degrees in a hexagon.

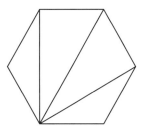

See if you can find the number of degrees in these polygons:

(a) Pentagon

(b) Heptagon

(c) Decagon

(d) Dodecagon

(e) Icosagon

TASK 9

3D objects

Recognising polyhedra and theorem

You will need:

- BLM 66

Explain to students that some shapes like spheres and cylinders have curved surfaces and others like prisms and pyramids have flat surfaces. Shapes with flat surfaces are called 'polyhedra', from the ancient Greek words 'poly' meaning 'many' and 'hedra' meaning 'bases' (faces). Polyhedra are formed by joining polygons together. When a polyhedron is pulled apart, the blueprint for its construction is called a 'net'. Types of polyhedra include cylinders, cubes, spheres, pyramids and rectangles.

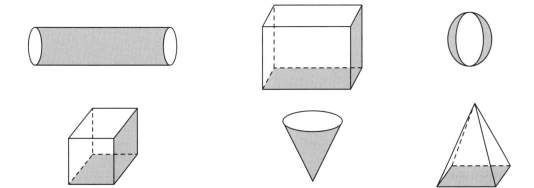

1. Let's go on a polyhedra hunt in your house and in your garden and list examples. Leonard Euler was a Swiss mathematician who discovered about 300 years ago that the properties of a polyhedron are always in a special balance. He found that the number of faces of a polyhedron when added to the number of its vertices (corners) will always equal 2 more than the number of its edges (F + V = E + 2).

2. Apply Euler's discovery (called Euler's Theorem) to the polyhedra listed on BLM 66.

Shape	Number of faces	Number of vertices	Number of edges	Apply $F + V = E + 2$
Cube				
Rectangular prism				
Triangular prism				
Square-based pyramid				
Tetrahedron				
Octahedron				
Pentagonal prism				
Pentagonal pyramid				

TASK 10

Going on a pi hunt

Finding parts of the circle

You will need:

- BLM 67
- calculator
- string
- ruler

Remind students that the middle of a circle is called the 'centre', the perimeter of a circle is called its 'circumference', any line that goes from the centre to the circumference is called a 'radius', and the straight line that goes across the circle through the centre is called the 'diameter'.

Parts of a circle

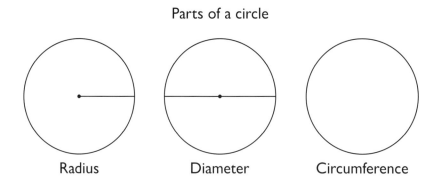

Radius   Diameter   Circumference

The ancient Greeks found that when you divide the length of the diameter of a circle into the length of the circumference of that circle, no matter how big or small the circle is, you always get the same answer, about 3 and a bit, which is usually rounded to two decimal places: 3.14. This

          OXFORD UNIVERSITY PRESS

relationship between the diameter and the circumference of a circle was given the name 'pi', from a letter of the Greek alphabet.

Find as many different circles that you can inside and outside your house. These could be the tops of cans, bread and butter plates, dinner plates, CDs, records, bin lids, etc. Then use:

- *a ruler to measure the diameter of each circle*
- *string and a ruler to measure the circumference of each circle*
- *a calculator to divide the diameter into the circumference of each circle*
- *a calculator to find how close to 3.14 were your answers.*

Type of circle	Diameter	Circumference	Circumference divided by diameter	Answer	How close to 3.14

TASK 11

Problems to solve

Applying geometry to unfamiliar situations

1. A bag contains rectangles, hexagons and 2 pentagons. Altogether there are 60 edges in the shapes in the bag. How many hexagons might be in the bag?

2. This compass has a needle that currently points to north. In a game, a move is described as: 'Go 45 degrees anticlockwise and then 90 degrees clockwise'. After 100 such 'moves', in which direction will the needle be pointing?

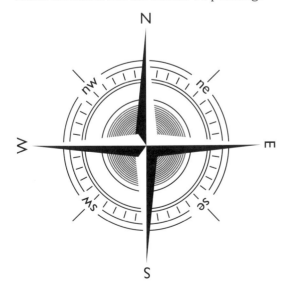

3. Sami plotted points on a grid at the co-ordinates (0,0) (1,2) (2,4) (3,6) (4,8) and (5,10), and joined them to form a straight line. How many degrees would this line be from the *x*-axis?

4. If you were to multiply together the edges, faces and vertices of the following 3D objects, which would give the highest product?

   (a) A cube.

   (b) A square-based pyramid.

   (c) A pentagonal prism.

5. Which 9 capital letters when reflected in a mirror and then rotated 180 degrees look exactly the same?

ANSWERS

Task 1 What does it mean?

1. Answers will vary but may include when: using a compass, using GPS, working with a grid, playing battleships, following a map, looking for something.

2. Answers will vary but may include when: looking at corners of things, looking at edges of things, walking up or down a hill, going up or down stairs, looking at the top of a building or a tree.

3. Answers will vary but may include when: looking at tiles, looking at the top of a table, looking at a clock face, looking at the front, top or back of any 3D object.

4. Answers will vary but may include when looking at a: can, box, packet, table, house, crystal, tube, wheel.

Task 2 Robots

Answers will vary.

Task 3 Architect and builder

Answers will vary.

Task 4 Compass points

Answers will vary.

Task 5 Angles all around

1. Answers will vary.

2. 19 degrees.

3. Answers will vary.

4. **(a)** E, F, H, L, T

   (b) A, K, M, N, V, W, X, Z

   (c) B, C, D, G, I, J, O, P, Q, R, S, U, Y.

OXFORD UNIVERSITY PRESS

Task 6 Mirror image

1. A, H, I, M, O, T, U, V, W, X, Y, 8.
2. **(a)** DAD: No.
   (b) LEVEL: No.
   (c) TOOT: Yes.
   (d) BOB: No.
   (e) HANNAH: No.
3. Answers will vary.

Task 7 Co-ordinates

1. (1,3), (3,7) and (5,3).
2. Answers will vary.

Task 8 2D shapes

1. Answers will vary.

2.

Vertices	Polygon name	Vertices	Polygon name	Vertices	Polygon name
3	Triangle	4	Quadrilateral	5	Pentagon
6	Hexagon	7	Heptagon	8	Octagon
9	Nonagon	10	Decagon	11	Hendecagon
12	Dodecagon	13	Tridecagon	14	Tetradecagon
15	Pentadecagon	16	Hexadecagon	17	Heptadecagon
18	Octadecagon	19	Enneadecagon	20	Icosagon

3. **(a)** Pentagon = 540 degrees
   (b) Heptagon = 900 degrees
   (c) Decagon = 1440 degrees
   (d) Dodecagon = 1800 degrees
   (e) Icosagon = 3240 degrees

Task 9 3D objects

1 Answers will vary but may include:

Polyhedra	Examples of polyhedra
Cylinders	cans of soup, Pringles, salami, downpipes, spray cans, toilet paper
Spheres	tennis balls, cricket balls, hockey balls, baseballs, basketballs, globes
Cubes	sugar cubes, bedside tissue boxes, food cartons, cardboard boxes, dice
Rectangular prisms	cereal packets, butter, match boxes, desks, tables, computers, filing cabinets
Others	trophies, ice-cream cones, door jambs, wigwams, ornaments

2.

Shape	Number of faces	Number of vertices	Number of edges	Apply F + V = E + 2
Cube	6	8	12	6 + 8 = 12 + 2
Rectangular prism	6	8	12	6 + 8 = 12 + 2
Triangular prism	5	6	9	5 + 6 = 9+ 2
Square-based pyramid	5	5	8	5 + 5 = 8 + 2
Tetrahedron	4	4	6	4 + 4 = 6 + 2
Octahedron	8	6	12	8 + 6 = 12 + 2
Pentagonal prism	7	10	15	7 + 10 = 15 + 2
Pentagonal pyramid	6	6	10	6 + 6 = 10 + 2

Task 10 Going on a pi hunt

Answers will vary.

Task 11 Problems to solve

1. Four answers work. There could be 1, 3, 5 or 7 hexagons, along with 11, 8, 5 or 2 rectangles.
2. The compass needle moves 45 degrees clockwise after every move. Therefore it will be back at north at the end of every 8 moves. It will be at north after 80, 88 and 96 moves. So it will be pointing south after 100 moves.
3. The plotted line would have a gradient or slope that would be $\frac{2}{3}$ towards the y-axis, so its angle would be 60 degrees.
4. **(a)** Cube: $12 \times 6 \times 8 = 576$
   (b) Square-based pyramid: $8 \times 5 \times 5 = 200$
   (c) Pentagonal prism: $15 \times 7 \times 10 = 1050$
5. B, C, D, E, H, I, K, O, X

OXFORD UNIVERSITY PRESS

UNIT 16 – CHANCES ARE ...
Probability

Probability has become a more important mathematical concept over time. We tend to relate this topic to games of chance involving dice, coins, spinners and cards. As adults, we know that there is often a nexus between problem gambling and a lack of mathematical understanding and this alone justifies the teaching of probability. But probability also permeates our lives beyond such games. In fact, every time we consider risk and reward we are dealing with probability. This might manifest itself in the consideration of taking out an insurance policy or something as banal as considering packing an umbrella when the weather forecast says a 25% chance of rain.

Your students will never be as engaged as when they roll 3 dice to try and make a total of 18 or try to toss a coin to obtain the result of 5 heads in a row. It is, of course, the analysis of the data gathered during such games of chance that is the essence of the exercise. Games of probability are rich in mathematics, invariably containing fractions, data, graphs, counting and problem solving – and a heap of fun and engagement in the process.

During your brainstorming session, focus on the vocabulary of probability as a continuum from 'impossible' to 'guaranteed', as well as on games of chance. And do not forget to check the Bureau of Meteorology website forecast for the chance of rain today!

TASK 1
Roll race
Observing and recording outcomes
You will need:

- BLM 68
- dice

Look at the grid on BLM 68. Roll a die and colour in a square above the number you have rolled. Use a different colour for each of the 6 numbers. Do this until one number has all its column coloured. Which number won the 'Roll race'?

Now try this a second time. Did the same number win?

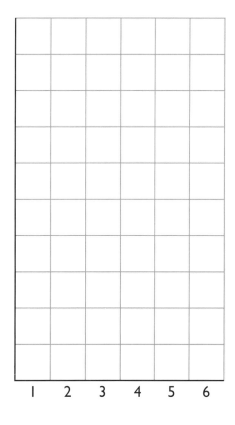

1	2	3	4	5	6

TASK 2

Eleven to 66

Exploring place value and probability

You will need:

- dice
- BLM 69

1. Roll a die 2 times to make a 2-digit number. The first roll will be the tens place and the second roll will be the ones place. So, if you roll a 3 then a 5 you will make 35.
2. Starting with 11 (made by rolling a 1 and then another 1) list all the possible 2-digit numbers you could make.

   Did you find all 36 different 2-digit numbers?
3. Look at the grid on BLM 69. In the blank spaces, finish filling in all 36 2-digit numbers that you could make.

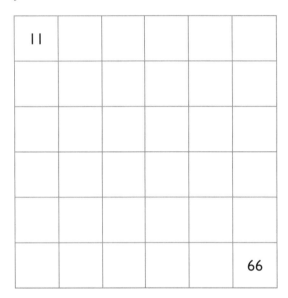

4. Now roll the die twie and colour in the 2-digit numbers that you make until the whole grid is full.
5. What was the first 2-digit number that you made twice?

TASK 3

Double trouble

Rolling doubles with 2 dice

You will need:

- dice

1. Roll a die twice:

   (a) Did you roll a double, like 3 then another 3? If not, keep rolling until you roll a double.

(b) How many rolls did it take?

(c) There are 6 possible doubles 1,1 2,2 3,3 4,4 5,5 6,6. Make a tally to see how many rolls it will take until you have rolled all 6 possible doubles.

2. Your task now is to try and roll a double 6:

(a) How many rolls do you predict it will take to roll a double 6? Make a tally.

(b) How many rolls did it take to roll the double 6?

(c) Try and roll a double 6 again. Did it take the same number of rolls?

TASK 4

Letter match

Exploring number names and probability

You will need:

- dice

1. **(a)** Roll a die and count the number of letters in the number that you roll. For example, if you roll a 3, 'three' has 5 letters in its number name. Now roll the die twice.

(b) Did the 2 numbers that you rolled have the same number of letters in their number names? If they did, like a 2 and then a 6, it is called a 'Letter match'.

(c) Now list all the possible ways of rolling 2 dice.

Remember that rolling a 1 then a 2 gives a completely different answer from rolling a 2 and then a 1.

2. How many of these rolls are a letter match?

3. **(a)** Therefore, what is the chance, as a fraction, of rolling 2 dice and getting a letter match?

(b) If you rolled the 2 dice 100 times, how many letter matches would you predict that you might roll? If you are brave enough you can roll 100 times to see how close your prediction is.

TASK 5

Coin flip

Tossing coins and predicting outcomes

You will need:

- a coin
- BLM 70

Remind students that many games of sport start with a coin toss. Encourage them to explore the mathematics behind tossing coins.

1. Get a coin and toss it 20 times. Record whether the coin landed heads up or tails up.

Toss	Heads or tails?	Toss	Heads or tails?	Toss	Heads or tails?	Toss	Heads or tails?	Toss	Heads or tails?
1		2		3		4		5	
6		7		8		9		10	
11		12		13		14		15	
16		17		18		19		20	

(a) How many heads did you toss?

(b) How many tails did you toss?

(c) Was this result as you had expected and why?

2. What do you think the chance is of tossing a head 5 times in a row? You can work this out by listing all the possible ways you could toss a coin 5 ways. The first way is HHHHH.

3. Now toss your coin to see how many attempts it will take to toss 5 heads in a row. Make a tally of your tosses.

4. How many tosses did it take?

5. Was this what you expected and why?

TASK 6

Come in spinner

Creating spinners and predicting outcomes

Spinners are a useful way to explore probability. In this activity students make their own spinner and make predictions about where it will land.

You will need:

- BLM 71
- cardboard
- sharp pencil
- paper clip

(a) Glue BLM 71 onto a piece of firm cardboard.

(b) Join the opposite corners of the hexagon with a straight line.

(c) Colour each of the 6 equal parts with a different colour.

(d) Write 1 to 6 on the sections.

(e) Push a sharpened pencil through the centre of the hexagon and the paper clip.

(f) Hold the pencil by its top and have the hexagon and paper clip level with a flat surface. Your spinner is ready to go!

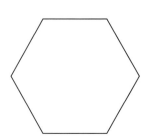

1. Spin your spinner 18 times and record your results in the table by using ticks:

Number	1	2	3	4	5	6
Ticks						
Total						

2. Rank your spins in order from number most spun to number least spun.

3. Was this the result that you had expected to happen? Why?

4. Now, let's make some predictions about what might happen if you were to spin your spinner a large number of times:

Times spun	1	2	3	4	5	6
6						
12						
24						
60						
300						
600						

TASK 7

On the deck

Applying probability to a deck of playing cards

You will need:

- playing cards

Remind students that many card games involve probability.

1. Give students a deck of playing cards and remove the jokers:

    (a) How many cards are in the deck?

    (b) How many cards are red?

    (c) How many cards are black?

    (d) How many cards are clubs or diamonds or hearts or spades?

    (e) How many different face cards are there, like ace or jack?

    (f) How many of each type of face card are there in the deck?

2. As a fraction, what is the chance of picking these cards?

    (a) A red card:

    (b) A 3

    (c) A spade

    (d) The 7 of hearts

    (e) A number card

    (f) A picture card

3. Now find a big surface and, after shuffling the cards, place them all face down.

   Select a card and pretend you are playing 'Fish'. You now need to find another card with the same face as the one in your hand. What is the chance, as a fraction, of this happening?

4. Place all the cards back face down on the flat surface. Pick a card. What is the chance that the card you picked will either be red *or* a 10?

5. What is the largest number of cards that you could pick before you would be guaranteed of picking a red card?

TASK 8

Dice roll

Exploring probability when rolling of 3 dice

You will need:

- 3 dice

1. Roll a die 3 times.

   Let's say that you have rolled 1, 6, 4. This has made the 3-digit number 164.

   List all possible 3-digit numbs that can be made this way, that are less than 211.

2. Now roll the die another 3 times creating a 3-digit number. Was it bigger than 166?

   (a) What do you think was the chance that it was bigger than 166?

   (b) If you rolled the 3 dice 30 times, how many rolls do you think would be bigger than 166?

3. Now roll the 3 dice 30 times and record your 30 3-digit numbers.

   How close was your prediction to your results?

TASK 9

Triple sums

Adding numbers rolled with 3 dice

You will need:

- 3 dice

1. Roll the 3 dice.

   If, for example, you have rolled 2, 5, 3 this is worth 10 points.

   A roll of 5, 2, 3 is a different roll worth 10 points and so is a roll of 6, 1, 3.

   (a) How many different rolls do you think will add up to 10?

   (b) Now list all the ways that you can roll a total of 10 with 3 dice.

2. Your task is now to try and make a total of 17 by rolling 3 dice. Keep a tally to see how long it will take:

   (a) How many rolls did it take?

   (b) This is normally way harder than making a total of 10. Why do you think this is true?

   (c) Now list all the ways that you can make a total of 17 by rolling 3 dice and adding up the numbers rolled.

TASK 10

Spinning around

Exploring probability

You will need:

- BLM 71

Encourage students to use their knowledge of chance and patterns to predict outcomes with spinners.

1. Look at the following spinner:

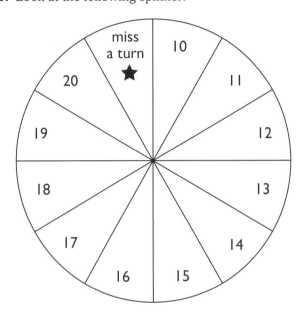

    (a) What is the chance of landing on any section?

    (b) What is the chance of landing on a number?

    (c) What is the chance of landing on a prime number?

    (d) What is the chance of landing on an odd number or the star?

    (e) What is the chance of landing on a perfect square?

2. If you spun the spinner 60 times, how often would you expect to spin the following?

    (a) 15.

    (b) A number smaller than 15.

    (c) A 2-digit number.

    (d) A number in the 3 or 4 times tables.

3. Using two copies of the hexagon on BLM 71, create 2 6-sided spinners. There are 2 possible answers to this task. The spinners have the following rules:

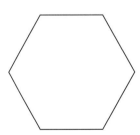

- There will be a chance of $\frac{1}{3}$ of landing on a 3.

- There will be a 50/50 chance of landing on an even number.

- There will be a $\frac{1}{6}$ chance of landing on a 2-digit number smaller than 14.

- There is a $\frac{2}{3}$ chance of landing on a number in the 3 times table.

- There is a $\frac{5}{6}$ chance of landing on a 1-digit number.

- The numbers on the spinner add up to 39.

TASK 11

Problems to solve

Applying probability skills to unfamiliar situations

You will need:

- dice
- BLM 71
- deck of cards

1. A spinner is numbered with the first 6 prime numbers. What would the chance be, as a fraction, of spinning the following?

    (a) A 2-digit number.

    (b) A number containing the digit 3.

    (c) A number with 2 vowels in the spelling of its number name.

    (d) A number in the 2, 3 or 5 times tables.

    (e) A composite number.

2. Here are the rules of the game of 'Up to 100':

    Play with a partner, alternating rolls of a dice. Add up the numbers rolled.

    The first player to reach exactly 100 wins.

    What would be the fewest number of rolls of the dice to get a winner?

3. In how many different ways could you roll a dice 6 times, add up the results and make a total of 34? (Note: 466666 and 664666 are regarded as different rolls.)

4. Here are the rules of the game of 'Split':

    Play in pairs.

    Player A rolls 2 dice and adds up the numbers rolled. The same player repeats this. This gives 2 totals or targets.

    For example, rolls of 4,1 and 6,4 will give targets of 5 and 10.

    To win, Player B must roll the 2 dice, add up the numbers rolled and split the target of 5 to 10: so, totals of 6, 7, 8 or 9 will win. Totals of 5 or 10 are regarded as a no result.

    What is the chance of Player B winning this game?

5. You shuffle a deck of cards and pick the ace of hearts. You put this card back. You then shuffle the cards and hand them to a friend who also picks the ace of hearts. What is the chance of this happening?

OXFORD UNIVERSITY PRESS

ANSWERS

Task 1 Roll race

Answers will vary but there is a $\frac{5}{6}$ chance that the winning number will be different for the 2 grids.

Task 2 Eleven to 66

1. Answers will vary.
2. 2-digit numbers: 11, 12, 13, 14, 15, 16, 21, 22, 23, 24, 25, 26, 31, 32, 33, 34, 35, 36, 41, 42, 43, 44, 45, 46, 51, 52, 53, 54, 55, 56, 61, 62, 63, 64, 65, 66.
3. Answers will vary.
4. Answers will vary.
5. Answers will vary.

Task 3 Double trouble

The chance of rolling a double 6 is $\frac{1}{36}$.

Task 4 Letter match

1. (c) 11, 12, 13, 14, 15, 16, 21, 22, 23, 24, 25, 26, 31, 32, 33, 34, 35, 36, 41, 42, 43, 44, 45, 46, 51, 52, 53, 54, 55, 56, 61, 62, 63, 64, 65, 66
2. 14 of these rolls create a letter match.
3. (a) The chance of creating a letter match is $\frac{14}{36}$ or $\frac{7}{18}$. (b) $\frac{7}{18}$ is less than 50%, so a sensible prediction would be about 35 to 45 times.

Task 5 Coin flip

1. (a) Answers will vary.
   (b) Answers will vary.
   (c) The expected result would be 10 heads and 10 tails.
2. Answers will vary. The 32 ways are: HHHHH, HTHHH, HHTHH, HHHTH, HHHHT, HTTHH, HTHTH, HTHHT, HHTTH, HHTHT, HHHTT, HTTTH, HHTTT, HTHTT, HTTHT, HTTTT, TTTTT, THTTT, TTHTT, TTTHT, TTTTH, THHTT, THTHT, THTTH, TTHHT, TTHTH, TTTHH, TTHHH, THTHH, THHTH, THHHT, THHHH.
3. Answers will vary.
4. Answers will vary.
5. The expected number of tosses would be 32.

Task 6 Come in spinner

1. Answers will vary.
2. Answers will vary.
3. The expected result is 3 of each number.

4. 6 spins = 1 of each possible number.

12 spins = 2 of each possible number.

24 spins = 4 of each possible number.

60 spins = 10 of each possible number.

300 spins = 50 of each possible number.

600 spins = 100 of each possible number.

Task 7 On the deck

1. (a) 52 **(b)** 26 **(c)** 26 **(d)** 13 **(e)** 13 **(f)** 4.

2. (a) $\frac{1}{2}$ **(b)** $\frac{1}{13}$ **(c)** $\frac{1}{4}$ **(d)** $\frac{1}{52}$ **(e)** $\frac{9}{13}$ **(f)** $\frac{3}{13}$.

3. $\frac{3}{51}$ or $\frac{1}{17}$

4. $\frac{7}{13}$

5. 26

Task 8 Dice roll

1. 111, 112, 113, 114, 115, 116, 121, 122, 123, 124, 125, 126, 131, 132, 133, 134, 135, 136, 141, 142, 143, 144, 145, 146, 151, 152, 153, 154, 155, 156, 161, 162, 163, 164, 165, 166

2. (a) The chance of rolling a number greater than 166 is 5 chances out of 6.

(b) If you rolled 30 times then you would expect about 24 rolls to be bigger than 166.

3. Answers will vary.

Task 9 Triple sums

1. (a) There are 27 ways of rolling 3 dice and getting a total of 10:

(b) 136, 163, 316, 361, 613, 631, 145, 154, 415, 451, 514, 541, 226, 262, 622, 235, 253, 325, 352, 523, 532, 244, 424, 442, 334, 343, 433

2. (a) The chance of rolling a total of 17 is $\frac{3}{216}$ or $\frac{1}{72}$.

(b) A total of 10 is 9 times easier because there are 9 times as many ways of making the total.

As there are 216 ways of rolling 3 dice (6 × 6 × 6) the chance of rolling a total of 10 is $\frac{27}{216}$ or $\frac{1}{8}$.

(c) Making a total of 17 is much harder because there are only 3 ways in which it can be done: 566, 656, 665.

Task 10 Spinning around

1. (a) $\frac{1}{12}$ **(b)** $\frac{11}{12}$ **(c)** $\frac{1}{3}$ **(d)** $\frac{1}{2}$ **(e)** $\frac{1}{12}$.

2. (a) 5 times **(b)** 25 times **(c)** 55 times **(d)** 25 times.

3. Answer 1: 3, 3, 6, 7, 8, 12.

Answer 2: 3, 3, 6, 8, 9, 10.

Task 11 Problems to solve

1   The spinner will contain the numbers 2, 3, 5, 7, 11 and 13.

(a) $\frac{2}{6}$ or $\frac{1}{3}$.

(b) $\frac{2}{6}$ or $\frac{1}{3}$.

(c) $\frac{3}{6}$ or $\frac{1}{2}$.

(d) $\frac{3}{6}$ or $\frac{1}{2}$.

(e) $\frac{6}{6}$ or 0.

2. Answer: 33 rolls. Player A could roll 6 16 times, then a 4. Meanwhile, player B will have rolled 16 times.

3. 21 ways: 466666, 646666, 664666, 666466, 666646, 666664, 556666, 565666, 566566, 566656, 566665, 655666, 656566, 656656, 656665, 665566, 665656, 665665, 666556, 666565, 666655.

4. There are 36 ways of rolling 2 dice.

In the game described, rolls of 1,5 2,4 3,3 4,2 5,1 1,6 2,5 3,4 4,3 5,2 6,1 2,6 3,5 4,4 5,3 6,2 3,6 4,5 5,4 and 6,3 will win. Thus, the chance of a 'split' is $\frac{20}{36}$ or $\frac{5}{9}$.

5. The chance is $\frac{1}{52} \times \frac{1}{52}$ or $\frac{1}{2704}$.

UNIT 17 – NOT A PROBLEM
Problem-solving strategies

Due to the rapidly changing nature of the world in which we live, the best thing we can do for our students is to encourage them to become creative and flexible thinkers – open to change, compassionate by nature and resilient when faced with difficulties. Problem solving is a wonderful way to achieve these goals.

By definition, problem solving is the application of an acquired skill, knowledge and concepts to an unfamiliar situation. For students to become successful problem solvers, they need to develop a deep understanding of mathematical concepts, and be persistent and resilient workers. It is true that each topic that we cover in mathematics should have problem solving embedded in it, but spending time teaching the problem-solving strategies found in this unit is invaluable.

In my own school, at every year level, we devote at least one week a year to the study of problem-solving strategies, with excellent results as a consequence. It is also worth noting that the strategies found in this unit are not just related to mathematics: they are applicable to all subjects.

During a brainstorming session on problem solving, ask your students what they think problem solving is and what makes a good problem solver. It is also an excellent teaching ploy to ask your students to create their own problem-solving questions to share with their peers.

TASK 1

What is problem solving?

Understanding problem solving

Remind students that problem solving is basically applying the knowledge, skills and concepts that people have learned and developed to a situation that is unfamiliar or different from what they normally see. But problem solving involves some creativity – it is finding a solution when there is no obvious answer. So 'John has $5, Sam has $10 and Jessica has $15. How much money did the 3 children have altogether?' is *not* problem solving because what needs to be done ($5+$10+$15) is obvious and does not need the application of creativity to find the answer of $30. However, 'In how many ways in our money system, using notes, could you make $30?' is problem solving because not only is the answer not obvious, but a strategy needs to be used to find the answer in a logical way.

Can you see which of these questions are examples of problem solving and which are not?

1. Caitlin shook hands with Bill and Greg. Harry shook hands with Joe and Chloe. How many handshakes took place altogether?

2. In how many ways can 6 people shake hands with each other?

3. Four people were in a group to have a photo taken. In how many ways could the 4 people be grouped together to have their photo taken?

4. Karl had 123 photos and Bella had 67 photos. How many more photos did Karl have than Bella?

5. James was making a vegetable patch that needed a border. He bought 36 m of timber for the border at a cost of $11.50 per metre. How much did the vegetable patch border cost James?

6. What is the largest area that can be made with 36 m of timber?

7. What fraction fits exactly halfway between one-half and one-third?

8. One half + one-third.

9. I spent $12 at the butcher and $45 at the supermarket. I had $15 left. How much money did I start with?

10. I spent half my money at the butcher and half of what was left at the supermarket. I had $13 left. How much money did I spend at the butcher?

TASK 2

Key words

Identifying important words

Explain to students that looking for the important words in a question is an essential prerequisite to being able to solve any problem. Simply put, if we do not know what we need to do, we cannot do it. Whenever we are given a maths problem to solve, we need to read the instructions carefully, sometimes more than once. We should underline or circle the key words to help focus our attention on the key elements of the question, which, in turn, will help us formulate a strategy to help us find an answer.

List the important words and numbers in each of these problem-solving questions:

1. Mae is going shopping. She looked in her purse and noticed that she had 15 bank notes that together added up to $395. What was most unusual was that in her purse Mae noticed that she had examples of every note in our money system. What might the notes be that were in Mae's purse?

2. Gianni and Elli work for the Paint People. They have been professional painters for over 12 years and have a great reputation for reliability and excellence. Gianni can paint a 10 m by 3 m wall in 20 minutes. Elli can paint the same sized wall in 30 minutes. If they were to work together on a 10 m by 3 m wall how long would it take Gianni and Elli to paint such a wall?

3. Arjun Kharwaja is learning archery from his uncle, Kabir. Kabir was in Australia's Olympic archery team at both the 2004 Athens games and the 2008 Beijing games, making the final on both occasions. Arjun is becoming very interested in archery and although he is only 7 years of age, he is in the Victorian under-10s team. In practice, Arjun shot 3 arrows at a target with 10 rings marked from 1 to 10 points. His 3 arrows score 26 points in total. In how many different ways could Arjun have scored 26 points with 3 arrows if the order of the scores does not matter (e.g. 10, 2, 6 is seen as being the same as 6, 2, 10)?

4. Sara is on a plane flying to Sydney to attend her sister's wedding. She is in the wedding party and is staying at the Wentworth Hotel on George Street near Circular Quay. Sara loves maths and sees opportunities everywhere to count and solve problems. Sara notices that on the plane there are 36 children. She also notices that half of the passengers on the plane are men and two-fifths are women. Sara also notices that the morning tea snack she had with her cup of coffee was made in Keilor Downs, the suburb in which she lives. How many people were on the plane?

TASK 3

Patterns and connections

Finding a pattern

Remind students that maths is full of patterns and connections. In fact, it is our ability to see patterns that makes us good mathematicians. By being able to see patterns we can often 'cut to the chase' in a problem and in so doing, save ourselves a huge amount of time. Ask students to find the pattern and then solve these problems:

1. What is the 234th odd number?
2. In how many ways can the digits 1, 2, 3, 4 and 5 be used to form 5-digit numbers in such a way that all 5 digits are used once?
3. A spinner is divided into 4 equal sections, clockwise numbered 1, 2, 3 and 4. Starting at 1, Zoe turns the needle a quarter turn anticlockwise 121 times. After 121 turns of the needle anticlockwise, to which number will the needle be pointing?
4. At the end of the basketball season the Antelopes, the Bears, the Cheetahs, the Dolphins and the Eagles made the finals and will play off in a round robin competition to find the premier team for the year. How many games will be played in the finals round robin series of matches?
5. At kindergarten, Georgia is stringing some beads together to form a pretty pattern. She follows the following rules: Every 3rd bead must be red. Every 4th bead must be yellow. When these numbers cross, like bead number 12, a green bead must be used. Every other bead must be blue. What colour will be bead 202 on the string?

TASK 4

Have a go

Attempting and checking answers

Students can often get stumped by a question and have difficulty, even if when they know what to do, to get 'out of the starting blocks'. This is where the 'Have a go' strategy comes into its own.

This strategy asks us to commit something to paper so that we can at least begin to try and solve the question. Sometimes this strategy is called 'Guess and check'. But the last thing we should ever do when we are dealing with a maths problem is 'guess'. The 'Have a go' strategy encourages students to think about what might possibly work, apply it to the question, test it and, if it does not work, alter their attempt, to try and get closer to an answer that *will* work.

'Have a go' at these problems. Test your answer and, if necessary, change your answer or strategy until you come up with an answer that works:

1. A farmer has 12 animals in his backyard. Some are pigs and some are chickens. The farmer counts 28 legs on his animals. How many are pigs?
2. When 14 is taken off a number it gives the same answer as when that number is divided by 3. What is that number?
3. Three numbers that are consecutive, add up to 144. What are the 3 numbers?
4. A number between 24 and 30 has its factors adding up to twice itself. What is that number?
5. Three times a number and 2 times a number that is 1 less than the first number have a difference of 13. What are the 2 numbers?

TASK 5

Record it

Using a table or a chart

You will need:

- BLM 72

Remind students that tables and charts are an excellent way of recording answers in a systematic, logical way. The best thing about a table or a chart is that it organises our thoughts in such a way that usually leads on to the recognition of trends and patterns.

Imagine that some good friends meet at a skate park. They have not seen each other for a long time so begin by greeting each other with a high five. If there are 20 such friends, how many high fives take place? The first three have been done for you.

Number of friends	Number of high fives	Number of friends	Number of high fives
1	0	11	
2	1	12	
3	3	13	
4		14	
5		15	
6		16	
7		17	
8		18	
9		19	
10		20	

What pattern did the use of this table help you to see?

TASK 6

Draw it

Using drawings and sketches

Remind students that drawing or sketching ideas can often help us understand what needs to be done to solve a problem, especially when combined with identifying key words in a question. Drawing is very useful for visual thinkers and helps to crystallise the context of the problem. Doing a drawing is often especially helpful in problems that involve measurement or geometry.

Ask students to use drawings or sketches to help solve these problems:

1. If a rectangle has a perimeter of 26 m and each edge is a whole number of metres in length, what is the greatest number of square metres that the area of this rectangle could have?

2. Draw a spinner that is split up into 6 equal sections marked clockwise, C, H, A, N, C and E. In a game, a move is described as: 'Put the spinner's needle on H and then move 2 places clockwise. Continue from the letter you landed on and keep moving 2 spaces clockwise'. If you do this 80 times, on which letter would you land?

3. A 4 cm × 4 cm × 4 cm cube is made by connecting 64 1 cm × 1 cm × 1 cm smaller cubes. It will then have its top and bottom faces painted red.
   How many of the 64 smaller cubes will not have any paint on them?

4. In a cross-country competition, a runner must run 1 km east, then 6 km north, then 1 km east, then 4 km south, then 3 km east, then 1 km north then 5 km west and finally 2 km south to the finish line.
   How far, in a straight line, is the start of the course from the finish line?

5. How big must a square be if its area is the same number of units as its perimeter?

TASK 7

Reverse it

Working backwards

Explain to students that 'working backwards' is a strategy that can be specific to a certain type of problem-solving question. These types of problems are typically long and contain a number of steps. Ask students to work backwards to solve these problems, and encourage them to check their answers, and try again if necessary:

1. I think that I must have lost some money because I know I left the house this morning to do some shopping and I had a $100 note in my wallet. I put $23 worth of petrol in the car, bought a hamburger and a juice costing $6 and $5 and then paid my newspaper bill which was $16. I then bought a $20 lottery ticket. When I came home I had $10 in my wallet.
   Am I correct in thinking that I have lost some money?

2. I took 7 from a number, doubled what was left, then added 6. I finished with 100. What number must I have started with?

3. Which number, when tripled and then halved, equals 24?

4. I spent half my money on a new book, half of what was left on bus fare and had $6.50 left when I came home. How much money did I have when I went into the bookshop?

5. I halved and halved and halved and halved and halved and halved and halved and halved and halved a number until I reached 4, my lucky number. At which number did I start the halving process?

TASK 8

Take it easy

Finding an easier way

Students often lose confidence or become intimidated by problems that appear very difficult to solve. However, many of these questions are nowhere near as hard as they may first seem. Encourage students to consider these super tough questions and try and alter the numbers in the question to make the solution more accessible. For example, 'What is a sensible answer to $\frac{11\ 1/17}{20} \times \frac{8\ 3/13}{13}$?' looks incredibly tough but when you round the mixed numbers to the nearest whole number, the problem becomes 12×8 and you know that the answer will be about 96 and that is all that the question wants you to do.

Try to make these problems easier by changing the numbers within them:

1. What is the answer to 123×999?
2. What is the ones digit in the answer when you multiply $7 \times 7 \times 7 \times 7 \times 7 \times 7 \times 7 \times 7 \times 7 \times 7 \times 7$?
3. What is the 41st number that is in all the 3, the 4 and the 5 times tables?
4. What is the closest number to 10 043 that is in the 3 times table?
5. What is the biggest 4-digit number that is in the 12 times table?

TASK 9

Model it

Making a model

This is another helpful strategy for visual thinkers and for helping to solve problems that involve measurement and geometry. Remind students that they can use paper, blocks or even people when they make models and they all work well.

Ask students to use suitable items to help solve these problems:

1. How many folds would it take for a piece of paper that is 24 cm long and 12 cm wide to end up with a perimeter of 18 cm?
2. Twenty-seven 1 cm × 1 cm × 1 cm cubes are connected to make a larger cube. The larger cube is then painted yellow on every face. How many of the 1 cm × 1 cm × 1 cm cubes that make up the larger cube will not have any paint on them?
3. Twenty-four individual and identical cubes are connected to make a rectangular prism. In how many different ways could this be done?
4. A sculpture, which consists of 3 cubes, sits on the ground in a park. The top cube is 1 m × 1 m × 1 m and sits on top, in the centre of a 2 m × 2 m × 2 m cube, which, in turn, sits on top and in the centre of a 3 m × 3 m × 3 m cube which touches the ground. The council then spray paints the exposed faces of the cube with blue paint.

   How much of the sculpture's surface area needs to be painted?

5. If the following 3D net was joined to form a solid, how many edges would it have?

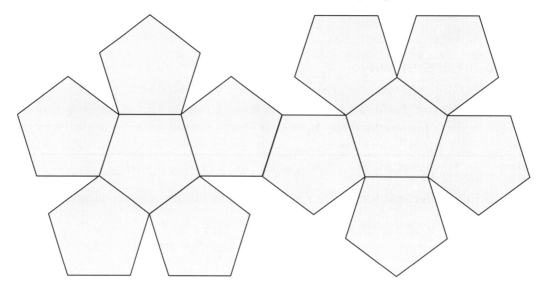

TASK 10

Think it through

Thinking logically

Logical thinking is about working through problems in a systematic manner. Many of the questions we encounter when problem solving have multiple parts or multiple answers. These questions require a systematic, logical approach, often requiring the use of prior knowledge to remove answers that are not possible, or listing answers in a sensible, numeric order.

Encourage students to apply a systematic approach, using logical thinking, to these problems:

1. What number am I? I am a 4-digit number. My digits sum to 20.

My number, when rounded to the nearest thousand, equals 5000. All my digits are different. My tens and ones digits have a product of 18.

I am the largest possible number to match all these clues.

2. There are 6 questions on a test named A, B, C, D, E and F. Each question can be marked out of 3, 2, 1 or 0, depending on the quality of the working out. In how many different ways could a student score 16 out of a possible 18 on such a test?

3. On a balance, 2 circles weigh the same as a triangle and 3 triangles weigh the same as a square. A rectangle weighs the same as 2 squares and a triangle. How many circles weigh the same as a rectangle?

4. In how many ways can you make 45c using cent coins in our money system?

5. At a closing down sale 3 T-shirts and 2 pairs of shorts cost $77, and 4 T-shirts and 1 pair of shorts cost $66. At the sale how much would 2 T-shirts and 3 pairs of shorts cost?

ANSWERS

Task 1 What is problem solving?

1. No.   **2.** Yes.   **3.** Yes.   **4.** No.   **5.** No.
6. Yes.   **7.** Yes.   **8.** No.   **9.** No.   **10.** Yes.

Task 2 Key words

1. Important words: 15 notes. $395. Examples of every note.

2. Important words: 20 minutes. 30 minutes. Together?

3. Important words: 3 arrows. Rings. 1 to 10 points. 26 points in total. Order does not matter.

4. Important words: 36. Half. Two-fifths.

Task 3 Patterns and connections

1. The 234th odd number is 467. The pattern that helps solve the problem is that odd numbers come before even numbers. The 234th even number must be 234 × 2 = 468, and 468 − 1 = 467.

2. The digits 1, 2, 3, 4 and 5 can form 120 different 5-digit numbers. The pattern that helps solve the problem is 5. There are 24 ways of forming 5-digit numbers starting with 1: 12345, 12354, 12435, 12453, 12534, 12543, 13245, 13254, 13425, 13452, 13524, 13542, 14235, 14253, 14325, 14352, 14523, 14532, 15234, 15243, 15324, 15342, 15423 and 15432. If 24 of these answers start with the digit 1, then it logically follows that 24 answers must start with each of the digits 2, 3, 4 and 5. 24 × 5 = 120.

3. It will be pointing to 4. The pattern that helps solve the problem is that the needle will be at 1 every 4 turns. Therefore it will be at 1 after 40, 80 and 120 turns. The 121st turn will have it back at 4.

4. There will be 15 finals games. The pattern that helps me solve the problem is 5 + 4 + 3 + 2 + 1.

5. The bead will be blue. The pattern that helps solve the problem is that the number 202 is not in the 3 times table (numbers in the 3 times table have their digits summing to 3, 6 or 9 and 2 + 0 + 2 = 4). The number 202 is also not in the 4 times table (numbers in the 4 times table have their last 2 digits divisible by 4, and 2 is not divisible by 4). So the bead must be blue.

Task 4 Have a go

1. The farmer must have 2 pigs and 10 chickens. All other combinations that may be tried give either too many legs or not enough animals.

2. 21 − 14 = 7 and 21 divided by 3 = 7.

3. The 3 numbers are: 47, 48 and 49. The easiest way of getting the answer is to divide 144 by 3. This will give you the middle of the three numbers.

4. The number 28 has factors of 1, 2, 4, 7, 14 and 28. 1 + 2 + 4 + 7 + 14 + 28 = 56.

5. 3 × 11 (33) and 2 × 10 (20) have a difference of 13.

Task 5 Record it

Number of friends	Number of high fives	Number of friends	Number of high fives
1	0	11	55
2	1	12	66
3	3	13	78
4	6	14	91
5	10	15	105
6	15	16	120
7	21	17	136
8	28	18	153
9	36	19	171
10	45	20	190

The pattern is that as a new person enters the group you add on 1 less than the number of people in the group.

20 people = 1 + 2 + 3 … + 19.

Task 6 Draw it

1. The closer the rectangle is to a square the greater will be the area. A 6 m × 7 m rectangle will have the greatest possible area of 42 sq m.
2. The spinner will land on E. Only H, N and E will be in the game. Every third landing will be H.

   80 = 78 + 2, so it will land on E.
3. The cube will contain 4 × 4 × 4 or 64 1 cm × 1 cm × 1 cm cubes. The top and bottom layers will both contain 16 + 16 or 32 1 cm × 1 cm × 1 cm cubes. Thus 32 will not have paint on them.
4. The finish line will be 1 km north of the start line.
5. The square must be 4 units on each edge: 4 + 4 + 4 + 4 = 4 × 4.

Task 7 Reverse it

1. Yes, you are correct. Your shopping only cost $70. You should have $30 left in your wallet, not $10.
2. When we work backwards we must reverse the operations that we are dealing with. So, to find the answer we need to create this equation: 100 − 6 ÷ 2 + 7. Thereby giving 54 as the solution.
3. If a number has been tripled and then halved to give 24 we need to reverse what has been done. Thus 24 needs to be doubled, then divided by 3. This gives us 48 ÷ 3 = 16.

OXFORD UNIVERSITY PRESS

4. I must have had $26 to spend because $6.50 is half of $13 and $13 is half of $26.

5. If you halve a number 9 times to reach 4, then 4 needs to double 9 times to find where you started. This gives 8, 16, 32, 64, 128, 256, 512, 1024, 2048.

Task 8 Take it easy

1. An easier problem would be 123×1000 which equals $123\,000$. If you take 123 away from this the answer is simply $122\,877$.

2. The question is only asking for the ones digit, therefore we only need to consider the last digit and look for a pattern.

   Following 7, $7 \times 7 = 49$, so 7 then 9 in the ones place so far.

   49×7 will give an answer with 3 in the ones place. So 7, 9, 3 so far.

   3×7 will end in a 1. So 7, 9, 3, 1, 7, 9, 3, 1, etc.

   The pattern must then continue in the ones place.

   The question is asking for the 11th number in the pattern: 7, 9, 3, 1, 7, 9, 3, 1, 7, 9, 3.

   The ones digit must be a 3.

3. All we need to do here is find the 1st number in the 3, 4 and 5 times tables, which is 60.

   $41 \times 60 = 2460$.

4. Look at the numbers in the 3 times table: 3, 6, 9, 12, 15, 18, 21, 24 … Can you see that each has its digits summing to either 3 or 6 or 9? The digits in $10\,043$ sum to 8. Add 1 more, $10\,044$, and you have a number whose digits sum to 9 and must be in the 3 times table.

5. Consider the first number in the 12 times table. It is in both the 3 times table and the 4 times table. So, as we have seen above, its digits must sum to 3, 6 or 9. Numbers in the 4 times table have their last 2 digits as a multiple of 4, such as $21\,648$.

   So the biggest 4-digit number whose digits sum to 3, 6 or 9 and whose last 2 digits are in the 4 times table is 9996.

Task 9 Model it

1. Making a piece of paper that has been cut to be 24 cm × 12 cm shows that it has a perimeter of 72 cm.

   One fold will make the paper 12 cm × 12 cm, with a perimeter of 48 cm. The next fold will make it a 12 cm × 6 cm rectangle, with a perimeter of 36 cm. The next fold will make it a 6 cm × 6 cm square, with a perimeter of 24 cm. The final fold makes it a 6 cm × 3 cm rectangle, with a perimeter of 18 cm.

   Thus, 4 folds are required.

2. The central cube will not be painted. Of the 27 cubes, 26 will be painted yellow.

3. The rectangular prism could have dimensions of $24 \times 1 \times 1$, $12 \times 2 \times 1$, $8 \times 3 \times 1$, $6 \times 2 \times 2$ or $4 \times 3 \times 2$, thus there are 5 possible rectangular prisms that could be made.

4. The top cube will have 5 sq m painted. The middle cube will have 19 sq m painted and the bottom cube will have 41 sq m painted. Altogether the cube will need to have 65 sq m painted.

5. This dodecahedron will have 30 edges.

Task 10 Think it through

1. A number of answers fit the clues: 4529, 4592, 4736, 4763, 5429, with 5492 being the largest.

2. There are 2 distinctly different ways that a student could score 16 out of 18: (a) losing 1 mark on 2 different questions, or (b) losing 2 marks on just 1 question. Thus the ways are:

   (a) Losing 1 mark on 2 different questions:

   223333, 232333, 233233, 233323, 233332, 322333, 323233, 323323, 323332, 332233, 332323, 332332, 333223, 333232, 333322 (15 ways).

   (b) Losing 2 marks on 1 question:

   133333, 313333, 331333, 333133, 333313, 333331 (6 ways).

   So there are 21 different ways of scoring 16 out of 18 on the test.

3. To solve this we need to equate everything to circles:

   1 triangle = 2 circles

   1 square = 3 triangles = 6 circles.

   1 rectangle = 2 squares and a triangle = 6 + 6 + 2 = 14 circles.

4. There are 9 ways.

   Most economical way to least economical way:

   20c + 20c +5c

   20c +10c + 10c + 5c

   20c +10c + 5c + 5c + 5c

   20c + 5c + 5c + 5c + 5c + 5c

   10c + 10c + 10c + 10c + 5c

   10c + 10c + 10c + 5c + 5c + 5c

   10c + 10c + 5c + 5c + 5c + 5c + 5c

   10c + 5c + 5c + 5c + 5c + 5c + 5c + 5c

   5c + 5c + 5c + 5c + 5c + 5c + 5c + 5c + 5c.

5. If we can get an equal number of shirts and shorts together, we can cancel out 1 of the items and work out the cost of the other:

   3 T-shirts + 2 pairs of shorts = $77

   4 T-shirts + 1 pair of shorts = $66

   If we double the bottom bill we get:

   3 T-shirts + 2 pairs of shorts = $77

   8 T-shirts + 2 pairs of shorts = $132

   This means that the difference between the 2 pairs of equations above gives us 5 T-shirts costing the difference between the 2 bills, $55.

   Therefore, a T-shirt must cost $11.

   If we put this fact back into the question then a pair of shorts must cost $22.

   Thus, the cost of 2 T-shirts and 3 pairs of shorts must equal $88.